UNDERSTANDING EXPLOSIONS

This book is one of a series of publications available from CCPS. A complete list of CCPS books appears at the end of this volume.

UNDERSTANDING EXPLOSIONS

Daniel A. Crowl

Department of Chemical Engineering
Michigan Technological University

Center for Chemical Process Safety
of the
American Institute of Chemical Engineers
3 Park Avenue, New York, NY 10016-5991

CCPS Publication G-61

Library of Congress Cataloging-in-Publication Data:
Crowl, Daniel A.
   Understanding explosions / Daniel A. Crowl.
      p. cm.
Includes bibliographical references and index.
   ISBN 0-8169-0779-X
   1. Chemical processes—Safety measures. 2. Explosions.
3. Combustion. I. Title.
   TP150.S24C764 2003
   660'.2804—dc21

                                         2003004849

This book is available at a special discount when ordered in bulk quantities. For information, contact the Center for Chemical Process Safety at the address shown above.

PRINTED IN THE UNITED STATES OF AMERICA
10 9 8 7 6 5 4 3 2 1

CONTENTS

3

PREVENTION AND MITIGATION OF EXPLOSIONS   113

Appendix D
PROCEDURE FOR EXAMPLE 3.2

Appendix E
COMBUSTION DATA FOR DUST CLOUDS

REFERENCES

GLOSSARY

INDEX

PREFACE

For over 40 years the American Institute of Chemical Engineers (AIChE) has been involved with process safety and loss control in the chemical, petrochemical, hydrocarbon process and related industries and facilities. The AIChE publications are information resources for the chemical engineering and other professions on the causes of process incidents and the means of preventing their occurrences and mitigating their consequences.

The Center for Chemical Process Safety (CCPS), a Directorate of the AIChE, was established in 1985 to develop and disseminate information for use in promoting the safe operation of chemical processes and facilities and the prevention of chemical process incidents. With the support and direction of its advisory and management boards, CCPS established a multifaceted program to address the need for process safety technology and management systems to reduce potential exposures to the public, the environment, personnel and facilities. This program entails the development, publication and dissemination of *Guidelines* relating to specific areas of process safety; organizing, convening and conducting seminars, symposia, training programs, and meetings on process safety-related matters; and cooperating with other organizations and institutions, internationally and domestically to promote process safety. Within the past several years CCPS extended its publication program to include a "Concept Series" of books. These books are focused on more specific topics than the longer, more comprehensive *Guidelines* series and are intended to complement them. With the issuance of this book, CCPS has published nearly 70 books.

CCPS activities are supported by the funding and technical expertise of over 80 corporations. Several government agencies and nonprofit and academic institutions participate in CCPS endeavors.

In 1989 CCPS published the landmark *Guidelines for the Technical Management of Chemical Process Safety*. This book presents a model for process safety management built on twelve distinct, essential and interrelated elements. The foreword to that book states:

> For the first time all the essential elements and components of a model of a technical management program have been assembled in one document. We believe the *Guidelines* provide the umbrella under which all other CCPS Technical Guidelines will be promulgated.

This Concept Series book *Understanding Explosions* supports several of the twelve elements of process safety enunciated in *Guidelines for the Technical Management of Chemical Process Safety* including process risk management, incident investigation, process knowledge and documentation, and enhancement of process safety knowledge.

In 1976, a monograph "Fundamentals of Fire and Explosion," authored by D. R. Stull of the Dow Chemical Company, was published as part of the AIChE Monograph Series (Volume 73, No. 10). Stull's work has long been out of print and no other publication has been available to replace it. AIChE and CCPS recognized the need for a similar, but updated, utilitarian work built upon the foundation provided by Stull and authorized the writing of this book.

The purpose of this book is to assist designers and operators of chemical facilities to better understand the causes of explosions so that they may design to prevent them and to mitigate their effects. This book should also prove useful for emergency response and homeland security and safety personnel.

ACKNOWLEDGMENTS

The American Institute of Chemical Engineers and the Center for Chemical Process Safety express their gratitude to all the members of the Vapor Cloud Explosions Subcommittee for their generous efforts and technical contributions in the preparation of this Concept Series Book.

Mr. John Davenport, formerly of Industrial Risk Insurers, and now a CCPS Staff Consultant, and Mr. Robert Linney of Air Products and Chemicals initially chaired the Vapor Cloud Explosions Subcommittee. Dr. Jan Windhorst of Nova Chemicals later assumed the chairmanship. The other subcommittee members were John V. Birtwistle, RRS Engineering PLC; William Thornberg, HSB Industrial Risk Insurers; Albert G. Dietz, Jr., U.S. Department of Energy; Alexi I. Dimopoulos, ExxonMobil Research & Engineering Company; Cheryl A. Grounds, Baker Engineering and Risk Consultants, Inc.; Peter Hoffman, Celanese Corporation; Randy Hawkins, Celanese Corporation, Robert A. Mancini, BP Amoco Corporation; Larry J. Moore, FM Global; Walter L. Frank, ABS Consulting; Ephraim Scheier, BP Amoco Corporation; Samuel A. Rodgers, Honeywell International Incorporated; David G. Clark, E. I. DuPont de Nemours & Company, and Gerald Meyers, U.S. Department of Energy. The contributions of David Kirby and R. F. Schwab are also acknowledged. Martin E. Gluckstein was the CCPS staff liaison and was responsible for the overall administration and coordination of the project.

The author is Professor Daniel A. Crowl of Michigan Technological University. Chad V. Mashuga drew almost all of the figures and assembled the tables while a graduate student at Michigan Technological University. Dr. Mashuga is currently employed by the BASF Corporation.

Before publication, all CCPS books are subjected to a thorough peer review process. CCPS also gratefully acknowledges the thoughtful comments and suggestions of the Peer Reviewers:

Quentin A. Baker, Baker Engineering and Risk Consultants, Inc.
Daniel Bourgeois, Proctor and Gamble Engineering
Rudy Frey, Kellogg Brown and Root
Stan Grossel, Process Safety and Design, Inc.
Shah Khajeh Najafi, SAFER Systems LLC

1

INTRODUCTION

Fires and explosions in the process industries, although rare, do occur and can cause loss of life, damage to the environment, loss of equipment and inventory, business interruption, and loss of public trust. Explosions may occur at fixed site facilities, and also during transportation. They can also occur in other industries besides the chemical industry, including food processing, utilities, pulp and paper, and pharmaceuticals, to name a few.

An analysis of the largest incidents in the chemical process industries, ranked by total capital losses shows that fires and explosions account for almost all of the major losses, as shown in the top of Figure 1.1. Figure 1.1 also shows the total capital losses over 5 year periods, adjusted to January 1998 dollars to account for inflation. These statistics show an upward trend in total accident costs. Clearly, fires and explosions are responsible for most of these large losses.

Damage from an explosion is caused by the resulting blast wave, thermal energy, flying fragments and debris, or the subsequent fire. Most of the damage due to explosions at a fixed site is usually limited to on-site effects. However, an explosion or fire can also result in a toxic release which can have additional off-site impacts. Toxic releases are discussed in more detail elsewhere (AIChE, 1989; Fthenakis, 1993; AIChE, 1996a; AIChE, 1997; AIChE, 1999a; AIChE, 2000).

The difference between a fire and an explosion depends on the time frame in which these events occur. Fires are typically much slower events involving the combustion of materials. Explosions are due to the sudden release of energy over a very short period of time and may or may not involve combustion or other chemical reactions. It is possible for a fire to lead to an explosion and an explosion might lead to a fire and secondary explosions if combustible gases or liquids are involved.

Most people understand what an explosion is, but a detailed technical definition is not simple, and many definitions are available. AIChE/CCPS (AIChE, 1994) defines an explosion as "a release of energy that causes a blast." They subsequently define a "blast" as "a transient change in the gas density, pressure, and velocity of the air surrounding an explosion point." Crowl and Louvar (Crowl and Louvar, 2002) define an explosion as "a rapid expansion of gases resulting in a rapidly moving pressure or shock wave." A further distinction for an explosion is that

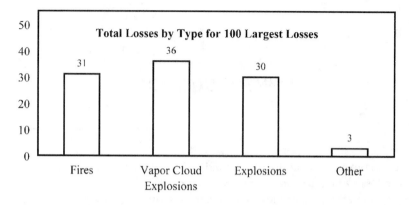

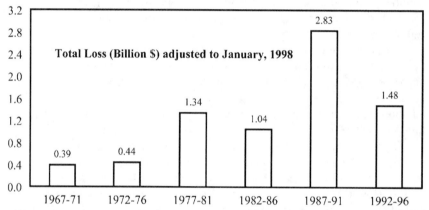

Figure 1.1. Hydrocarbon-chemical plant loss trends. (M & M Protection Consultants, 1998.)

the pressure or shock wave is of sufficient magnitude to cause potential damage or injury.

In summary, the three essential characteristics of an explosion are:

1. Sudden energy release
2. Rapidly moving blast or shock wave
3. Blast magnitude large enough to be potentially hazardous

This book deals with explosions. The purpose of the book is to provide information to those involved with the design, operation, maintenance and management of chemical processes. After reading this book the reader should understand the nature of explosions and the practical methods required to prevent explosions and to protect against their consequences. This book provides only limited detail on current explosion theories or models—many books and journal articles are available (see the References).

Anyone with a background in engineering, chemistry or a related technology should be able to read and apply this book. In particular, chemical plant operators, plant maintenance personnel, process engineers, design engineers, and plant management personnel should find this book useful. Others involved with first response and homeland safety and security could also use this book.

This book will provide the reader with an understanding of:

- The fundamental basis for explosions, from a practical standpoint.
- How to characterize the explosive and flammable behavior of materials.
- The different types of explosions.
- Hazard recognition related to fires and explosions.
- The practical methods to prevent an explosion or minimize the probability or consequence of an explosion during the routine use of flammable, combustible, and/or reactive materials.
- References for additional study.

This book has two parts. The first part (Chapter 2) describes the fundamentals of explosions. This includes the parameters used to characterize flammability and explosions of different materials, the various types of fires and explosions, and how explosions and fires cause damage. Examples are provided to show how the fundamentals relate to real-world problems. The second part (Chapter 3) describes the common, practical methods used to prevent explosions. Examples are provided here to demonstrate application of these methods.

1.1. Accident Loss History

In 1998 (USBL, 1999) there were 6010 accidental deaths on the job in the U.S. for all manufacturing sectors and job types. These fatalities are summarized in Table 1.1 according to cause. The leading cause of death was by transportation accidents, which were responsible for 44% of all fatal work injuries. Fires and explosions accounted for only 3% of the total fatalities.

These statistics might appear to indicate that fires and explosions are a minor risk in the workplace, but this is not the case. Risk is defined as "a measure of human injury, environmental damage or economic loss in terms of both the incident liklihood and the magnitude of the injury, damage or loss" (AIChE, 2000). Thus, risk is composed of both probability and consequence. For fires and explosions, the probability is low, but the potential consequence is high. In addition, fires and explosions are typically dramatic events, attracting wide media coverage, and harming the public's confidence in the industry. Figure 1.1 clearly shows that the consequences of fires and explosions are very high with respect to total dollar losses.

TABLE 1.1
Summary of Work-Related Fatalities By Cause for 1998 (USBL, 1999)

Cause	Total	Percent
Transportation accidents	2630	44
Homicide/suicide	960	16
Contact with objects and equipment	941	16
Falls	702	12
Exposure to harmful substances and environments	572	9
Fires and Explosions	205	3
Total fatalities	6010	100

1.2. The Accident Process (AIChE, 2000)

Accidents begin with an **incident**, which usually results in loss of control of material or energy. The incident could be the rupture of a tank by a forklift, a thermal runaway reaction, a leak in a flange due to corrosion of the connecting bolts, and so forth. A **scenario** describes the sequence of events that lead to the final consequence(s), or incident outcome(s), of the accident. For example, consider the incident of a tank failure resulting in the sudden loss of containment of a flammable liquid. Subsequent dispersion and mixing with air to form a flammable mixture may be followed by fire or explosion. The resulting explosion produces an **incident outcome** of a blast wave which causes damage, or **accident effects** on the surroundings.

1.3. A Case History—Flixborough, England

The Flixborough accident of June of 1974 will be used to demonstrate the steps in the accident process. This accident is selected here since it was extensively investigated, an inquiry with a detailed report was completed (Parker, 1975) and it involved the release and explosion of a large amount of flammable vapor. Many discussions of the accident are available elsewhere (Lees, 1986, 1996)

The Flixborough plant was designed to manufacture caprolactam, a precursor in the production of nylon. In one part of the process, cyclohexane is oxidized to a mixture of cyclohexanone and cyclohexanol. This reaction occurred in a series of six reactors shown in Figure 1.2. The process was maintained at a temperature of 155°C and a pressure of 7.9 atm to prevent the liquid from boiling.

The liquid inventory in the reactor system was large since the conversion and yield were low. The total cyclohexane liquid holdup in the reaction train was 120

Temporary Pipe Section

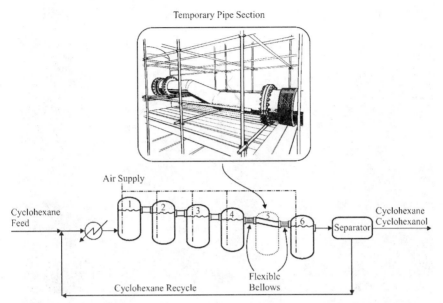

Figure 1-2. *The immediate cause of the Flixborough accident was the failure of the temporary pipe section replacing reactor 5. The illustration of the pipe rack is modified from Parker (1975). (Courtesy of HM Stationery Office.)*

metric tonnes. The lines between each vessel were originally 28 inches in diameter, with the liquid flowing by gravity from one vessel to the next.

About a month before the incident, a vertical crack was found in reactor 5. In order to maintain production the reactor was removed and a temporary bypass line was installed. The temporary, Z-shaped line was constructed of 20-inch diameter pipe with a flexible bellows at each end connecting reactors 4 and 6. The entire assembly was supported on a temporary scaffold, as shown in the cutout of Figure 1.2. This temporary design was sketched on a workroom floor and was not reviewed by a qualified engineer.

It is hypothesized that the temporary pipe assembly failed primarily due to repeated flexing of the bellows from process pressure changes causing the pipe to rotate somewhat due to its shape. The pipe assembly was also inadequately supported. The process was depressurized almost instantly, resulting in the release and vaporization of an estimated 30 metric tonnes of flammable liquid. The vapor mixed with the surrounding air and, about 45 seconds after the release, the vapor cloud was ignited, leading to the explosion.

The resulting blast killed 28 people, injured 36 other plant personnel, and destroyed the entire plant, including the administration building. Over 1800 nearby homes and 167 shops and factories were damaged. Fifty-three civilians

Figure 1-3. Damage in the reactor area from the Flixborough vapor cloud explosion.

were injured. Most of the fatalities occurred in the control room when the roof of the building collapsed. The accident occurred during a weekend—if it had occurred during normal working hours the administration building would have been filled with people and the fatality count would most likely have been greater. Figure 1.3 shows the resulting damage in the reactor area due to the explosion.

The subsequent investigation (Parker, 1975) found the following deficiencies:

*   The design of the temporary piping section was substandard and did not meet the design specifications of the original process.
*   A safety review or hazards analysis was not performed on the temporary piping section, as would be required today by management of change.
*   The plant site contained excessive inventories of flammable materials, which contributed to the accident after the initial blast.

1.4. Hazard Identification and Evaluation

A **hazard** is "an inherent physical or chemical characteristic of a material, system, process or plant that has the potential for causing harm" (AIChE, 1992a). This includes hazards associated with temperature, pressure, flammability, toxicity, etc. Many hazards are fixed and continuously present, such as chemical toxicity, but others occur due to process procedures and process conditions (such as high pressure).

An accident results when an incident occurs to activate a hazard. The incident could involve an operating or maintenance procedure, software, a material defect, corrosion, etc. More details on how this occurs are found elsewhere (AIChE, 1992b). In the Flixborough accident, the failure of the temporary pipe section activated the hazard associated with the flammable properties of the chemicals.

The purpose of **hazard identification** is to determine the hazards. Until a hazard is identified, it cannot be removed, controlled or mitigated. Many accidents, such as Flixborough, occur as a result of improper hazards identification.

Hazard evaluation is defined as "the analysis of the significance of hazardous situations associated with a process or activity." This includes a number of qualitative methods which are discussed fully elsewhere (AIChE, 1992a).

1.5. Inherently Safer Design

A chemical process is considered inherently safer if "it reduces or eliminates the hazards associated with materials and operations used in the process, and this reduction or elimination is permanent and inseparable" (Bollinger, Clark et al., 1996).

Inherently safer concepts can be applied at any point in the life cycle of a process, from conceptual design to process decommissioning. However, the largest benefits are realized when inherently safer design principles are applied during the early stages.

More information on inherently safer concepts is provided in Chapter 3 (Section 3.2) and elsewhere (Bollinger, Clark et al., 1996).

2

FUNDAMENTALS OF FIRES AND EXPLOSIONS

The explosive behavior of a material depends on many variables, including its physical state (solid, liquid or gas; powder or mist), its physical properties (heat capacity, vapor pressure, heat of combustion, etc.) and its reactivity. The type of fire or explosion that results also depends on a number of factors, including

- the material's initial conditions of use or storage
- the way in which the material is released
- how the material is dispersed and mixed with air
- when and how the material is ignited.

Figure 2.1 shows the major classifications of explosions. It is possible for more than one classification type to occur in any particular incident. These classifications will be discussed in more detail later in this text. Table 2.1 shows specific examples of the different types.

A **physical explosion** occurs due to the sudden release of mechanical energy, such as by releasing a compressed gas, and does not involve a chemical reaction. Physical explosions include vessel ruptures, boiling liquid expanding vapor explosions (BLEVE) and rapid phase transition explosions. The mechanical energy contained by the material in the vessel is released. A **vessel rupture** explosion occurs when a process vessel containing a pressurized material fails suddenly. The failure can be due to a number of mechanisms, including mechanical failure, corrosion, heat exposure, cyclical failure, etc. A **BLEVE** occurs when a vessel containing a liquified gas stored above its normal boiling point fails catastrophically. The vessel failure results in sudden flashing of the liquid into vapor, with subsequent damage due to the rapidly expanding vapor, ejection of liquid and vessel contents and fragment impact. A fireball may result if the material is combustible. A **rapid phase transition explosion** occurs when a material is exposed to a heat source, causing a rapid phase change and resulting change in material volume. A

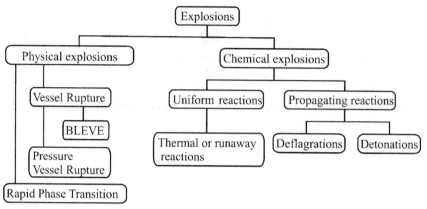

Damage effects due to: blast wave, thermal and/or projectiles.

Figure 2.1. *Relationship between the different types of explosions. It is possible for several to occur with any incident.*

TABLE 2.1
Examples of Various Types of Explosions

Type of Explosion	Examples
Rapid phase transition:	• Hot oil pumped into vessel containing water. • Valve in pipeline opened, exposing water to hot oil.
BLEVE:	• Corrosive failure of a hot water heater. • Propane tank rupture.
Vessel Rupture:	• Mechanical failure of a vessel containing high pressure gas. • Overpressuring of a vessel containing a gas. • Failure of a relief device during overpressure.
Uniform reaction:	• Thermal runaway of a continuous stirred tank reactor.
Propagating reaction:	• Combustion of flammable vapors in a fuel tank. • Combustion of flammable vapors in a pipeline.

chemical explosion requires a chemical reaction, which could be a combustion reaction, a decomposition reaction, or some other rapid exothermic reaction. A **uniform reaction** is a reaction that occurs uniformly through space in a reaction mass, such as a reaction which occurs in a continuous stirred tank reactor (CSTR). An example of an explosion caused by this type of reaction is the **runaway reaction** or **thermal runaway**. A runaway reaction occurs when the heat released by the reaction exceeds the heat removal, resulting in a temperature and pressure increase which may be sufficient to rupture the process containment. A **propagat-**

ing reaction is a reaction which propagates spatially through the reaction mass, such as the combustion of a flammable vapor in a pipeline, a vapor cloud explosion, or the decomposition of an unstable solid. Propagating reactions are further classified as **detonations** or **deflagrations**, depending on the speed at which the reaction front propagates through the unreacted mass. For detonations, the reaction front moves equal to or faster than the speed of sound in the unreacted medium and for deflagrations it moves at a speed less than the speed of sound.

Chemical explosions can occur in either the vapor, liquid, or solid phases. Chemical explosions which occur in the liquid or solid phases are sometimes called **condensed phase explosions**. These are significant due to the high energy density in the materials.

The damage from a fire or explosion is due to a number of impact mechanisms. This includes pressure effects, thermal exposure, projectiles and loss of material containment. For explosions, pressure effects are the most common. Any combination of these impacts is possible based on the particulars of the accident.

The following accident descriptions will demonstrate the relationships between the various explosion types:

Accident No. 1

An exothermic reaction was conducted in a large batch reactor. The reaction temperature was controlled by cooling water, which was supplied through coils within the reactor. The reactants were all flammable. The normal operating temperature of the reactor was above the normal boiling point of the reactants, so the vessel was pressurized to keep the reactants in the liquid state.

Due to an operational problem, the cooling water supply was interrupted, resulting in an increase in temperature within the reactor. This caused the reaction rate to increase and more heat to be generated. The result was a runaway reaction, causing the temperature within the reactor to increase very rapidly.

The pressure within the reactor also increased due to the increased vapor pressure of the liquids and the generation of gaseous products due to a decomposition reaction. The reactor vessel was equipped with a spring operated relief device, but the relief was undersized. Eventually, the pressure increased beyond the failure pressure of the vessel and it burst catastrophically.

Most of the liquid within the reactor almost immediately flashed to vapor due to the decrease in pressure. The vapor mixed with air forming a flammable mixture. The hot surface of a bare lightbulb in an adjacent area provided an ignition source, resulting in a vapor cloud explosion.

In this accident, three events occurred: a thermal runaway that developed sufficient pressure to burst the reactor, causing a physical explosion by rupture of the vessel and the flashing of the liquid, and a deflagration from the propagating combustion of the vapor cloud. Damage to the surroundings was caused by the blast wave from the initial vessel burst, followed by blast wave, thermal and projectile damage from the vapor cloud explosion. The initial vessel burst may also have caused projectile damage.

Accident No. 2 (Clayton and Griffin, 1994)

A large storage vessel in a juice and packing plant contained 66,000 lb of liquified carbon dioxide. The vessel was equipped with an internal heating coil and a relief device. Several weeks before the accident, the heating coils failed in the off state, resulting in a very low temperature excursion of the vessel.

Without warning, and without activation of the relief device, the vessel failed catastrophically, destroying the plant and killing three employees. The vessel pieces rocketed more than 1000 feet into an adjacent river.

An investigation concluded that on the day of the accident the heating coils apparently failed on, resulting in overheating and overpressuring of the vessel. It was conjectured that the relief device did not operate since it was plugged with ice—this had been observed in other installations. It was further concluded that a poor-quality weld had been weakened by the previous low temperature excursion and the catastrophic failure was initiated at this weld.

This accident involved the physical explosion of a nonreactive material. The damage from this accident was caused by the shock wave from the failing vessel followed by rapid flashing and expansion of the carbon dioxide liquid.

Accident No. 3 (USCSHIB, 1998)

An 18,000 gallon storage tank supplied propane for a turkey farm. At approximately 11 pm two youths were riding an all terrain vehicle (ATV) and struck both the vapor and liquid lines from the tank. The liquid line was completely severed from the tank at a location where it was connected to a manual shut-off valve directly beneath the tank. An excess flow valve connected to the liquid line failed to function since the line size was too small to provide adequate flow to activate it. Escaping propane from the damaged lines formed a vapor cloud and ignition occurred a few minutes after the line rupture. The likely ignition source was a propane vaporizer about 40 feet away from the tank.

The volunteer fire department responded and arrived on the scene at 11:21 p.m. The plan was to let the fire burn itself out and at the same time to water down the adjacent buildings to prevent the spread of the fire. At approximately 11:28 the tank exploded. Two fireman located 100 feet away from the tank were killed by a large tank fragment. Seven others were injured. The tank and associated piping fragmented into at least 36 pieces.

The investigation concluded that the vapor and liquid lines were inadequately protected, the liquid line was inadequately sized to activate the excess flow valve, and the emergency response procedure by the volunteer fire department did not provide an adequate hazard zone around the tank.

This is an example of a physical explosion involving a flammable material resulting in a classic BLEVE. The fire exposure weakened the walls of the vessel and caused it to fail catastrophically. Damage in this incident was due primarily to the resulting fragments.

2.1. Gases and Vapors

Figure 2.2 shows the fire triangle describing the requirements for a fire of a gas or vapor. The three requirements are fuel, oxidant and an ignition source. If all three components are present, a fire/explosion could result. If any of the three components are removed, then a fire or explosion is not possible.

The most common oxidant for a fire or explosion is air. Other materials can provide oxygen, such as hydrogen peroxide, perchloric acid, ammonium nitrate and metal and organic peroxides. Other materials can take the role of an oxidizer, including chlorine and fluorine. Exothermic decomposition, without oxygen, is also possible, for example, with ethylene oxide or acetylene.

In the past, fire protection focused primarily on removal of the ignition source. Practical experience has shown that this approach is not robust enough. Ignition sources are present almost everywhere, and it takes very little energy to ignite a flammable gas or vapor. The current practice to prevent fires and explosions is to continue with the elimination of ignition sources, while focusing efforts more strongly on preventing flammable mixtures.

The 20-liter laboratory vessel shown in Figure 2.3 is a typical device used to characterize the combustion behavior of gases, vapors and dusts. The vessel is first evacuated and then a test sample is introduced. The vessel is equipped with a high speed pressure transducer to track the pressure during the combustion process. The

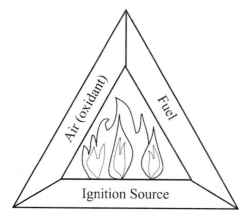

Figure 2.2. The fire triangle showing the requirement for combustion of gases and vapors.

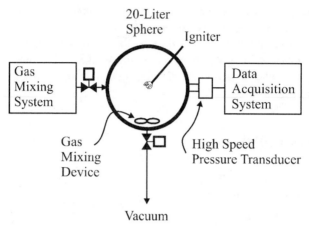

Figure 2.3. *An apparatus for collecting explosion data for gases and vapors.*

mixture is ignited by an igniter device in the center of the vessel. The ignition can be achieved by a spark, an exploding fuse wire, a hot wire or a chemical igniter.

Figure 2.4 is a plot of the pressure versus time behavior of this burning gas. In this case the gas is methane—although this behavior is typical for most flammable gases and vapors. After the initial ignition the pressure rises rapidly to a maximum

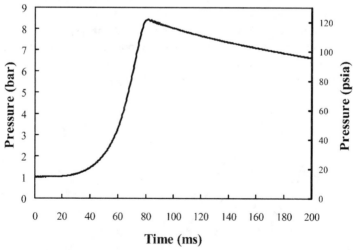

Figure 2.4. *Pressure versus time history for the explosion of a mixture composed of 10% methane, 70% nitrogen, and 20% oxygen in a 20-liter test sphere. The initial temperature and pressure is 25°C and 1 atm (Mashuga and Crowl, 1998).*

pressure, followed by a much slower decrease in pressure as the combustion products are quenched and cooled.

The combustion process demonstrated by the pressure history on Figure 2.4 is called a **deflagration**. A deflagration is a combustion process where the reaction front moves at a speed less than the speed of sound in the unreacted gases. The reaction front for a deflagration is propagated primarily by conduction and diffusion of the energy and free radical species into the unreacted gas. This is distinct from a **detonation**, where the reaction front moves faster than the speed of sound in the unreacted gases. For a detonation, the reaction front is propagated primarily by compressive heating of the unreacted gases ahead of the reaction front. The technical difference between a deflagration and detonation might appear small. However, the two types of explosions have very different behaviors, with outcomes that may or may not be the same. These differences will be discussed in more detail in Section 2.7 on Gas Dynamics.

Deflagrations are more common than detonations in explosions of flammable vapors or gases. A deflagration to detonation transition (DDT) is possible under certain circumstances (see Section 2.7).

Figure 2.5 shows two parameters that are used to characterize the explosive behavior for deflagrative explosions. The first variable is the maximum pressure during the unvented combustion, P_{max}. The second variable is the maximum slope of the pressure curve, $(dP/dt)_{max}$. Studies have shown that the maximum explosion pressure remains essentially constant, but the maximum slope of the pressure curve (i.e., pressure rate) decreases as the test volume increases, all other factors

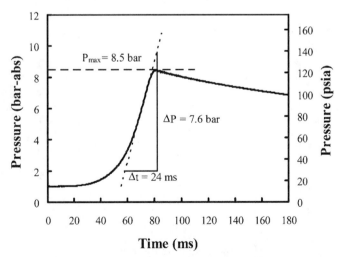

Figure 2.5. *The maximum pressure, P_{max}, and the maximum pressure rate are used to characterize the combustion data of Figure 2.4. These data were determined using a 20-liter sphere (Mashuga and Crowl, 1998).*

being held constant. The change in pressure rate is represented by the following empirical formula (Bartknecht, 1981).

$$K_G = \left(\frac{dP}{dt}\right)_{max} V^{1/3}$$
(2-1)

The parameter K_G is called the **deflagration index**. The combustion is considered more violent the higher the value of the deflagration index.

The deflagration index is computed from the data from Figure 2.5 for this oxygen enriched methane mixture as follows:

$$K_G = \left(\frac{dP}{dt}\right)_{max} V^{1/3} = \left(\frac{7.6 \text{ bar}}{0.024 \text{ s}}\right)(0.020 \text{ m}^3)^{1/3} = 86.0 \text{ bar m/s}$$

The maximum pressure, P_{max}, and the deflagration index, K_G, are important in characterizing the behavior of the combustion and in designing protection systems. Typical values are shown in Table 2.2. The maximum pressure values are fairly consistent between researchers. However, there is wide variability in the deflagration indices reported by different investigators, most likely the result of sensitivity to gas composition, humidity, ignition strength, etc. (Mashuga and Crowl, 1998). Neither the maximum pressure nor the deflagration index are inherent physical properties of the chemical, but are important quantities derived from specific experimental procedures and conditions.

Figure 2.6 is a plot of the maximum pressure, P_{max}, as a function of methane concentration in air. As the methane concentration decreases, a point is reached

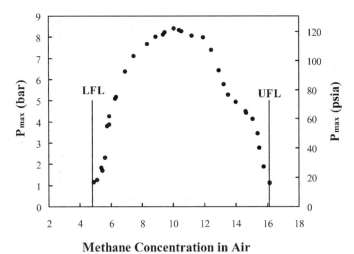

Methane Concentration in Air

Figure 2.6. Maximum pressure as a function of volume percent concentration for methane in air in a 20-liter test sphere. The initial temperature and pressure is 25°C and 1 atm (Mashuga and Crowl, 1998). The stoichiometric concentration is 9.51% methane.

where the concentration is too lean to support combustion. This is called the **lower flammability limit** (LFL). As the methane concentration is increased, a point is reached where the mixture is too rich in fuel to support combustion. This is called the **upper flammability limit** (UFL). The upper and lower flammability limits are also called the upper and lower explosion limits, UEL and LEL. Note that the flammability limits are normally defined with respect to fuel mixtures with air, but they can also be defined for other gas mixtures, such as with oxygen enriched air or even pure oxygen. Flammability limits will differ widely in enriched oxygen atmospheres or with other oxidants such as chlorine. Other inert gases besides nitrogen, such as carbon dioxide or argon, will affect the flammability limits. The limits are normally also specified for ambient temperature and pressure. Increased temperature or pressure will typically expand the flammability limits—some mixtures which are not flammable at ambient temperature and pressure may become flammable at increased temperature or pressure.

A reference is sometimes made to a **detonable limit**, which is the limiting concentration at which a detonation can occur in a gas mixture. These limits are frequently confused with explosion limits. The detonable limits are not well-defined and are highly sensitive to the specific apparatus and ignition source and are not routinely used.

From a technical standpoint, it is difficult to provide a precise definition to a flammability limit. From a practical standpoint, the American Society for Testing and Materials (ASTM) defines a fuel mixture with air as *flammable* if the pressure increase during the combustion process is more than 7% of the initial pressure, as described in ASTM E918 (1992). Mixtures that produce smaller pressure increases

TABLE 2.2
Maximum Pressures and Deflagration Indices for a Number of Gases and Vapors

	Maximum Pressure P_{max} (barg)			Deflagration Index K_G (bar-m/sec)		
Chemical	NFPA 68 1998	Bartknecht 1993	Senecal 1998	NFPA 68 1998	Bartknecht 1993	Senecal 1998
Hydrogen	6.9	6.8	6.5	659	550	638
Methane	7.05	7.1	6.7	64	55	46
Ethane	7.8	7.8	7.4	106	106	78
Butane	8.0	8.0		92	92	
Isobutane			7.4			67
Propane	7.9	7.9	7.2	96	100	76
Pentane	7.65	7.8		104	104	
Ethylene			8.0			171
Methyl Alcohol		7.5	7.2		75	94
Ethyl Alcohol			7.0			78
Ethyl benzene	6.6	7.4		94	96	

(or none at all) are called *not flammable*. The flammability limits are defined at the fuel concentrations where the pressure increase is exactly 7%. Other definitions and methods (Checkel, Ting et al., 1995) to determine the flammability limits are available. Flammability limits for a number of common materials are provided in Appendix C. Flammability limit data are normally available on material safety data sheets (MSDS) provided by the chemical manufacturer. An excellent summary of flammability limits is provided by Britton (2002a).

Another difficulty in characterizing flammability is that the parameters (flammability limits, deflagration index and all other parameters) are not fundamentally based. They are actually phenomenological descriptors based on a particular experimental apparatus and procedure. Thus, the experimental apparatus and procedure frequently becomes the issue.

Other quantities sometimes reported for gas or vapor combustion are the **flame speed, fundamental burning velocity, burning velocity, laminar burning velocity** and **turbulent burning velocity**. Wide variability exists in the definitions and usage of these terms.

Glassman (1996) defines the **flame speed** as the speed at which the combustion wave moves relative to the unburned gases in the direction normal to the wave surface. He states that this is also called the burning velocity, normal combustion velocity, or flame velocity. He also describes several experimental procedures to determine flame speeds.

NFPA 68 (1998) defines the **fundamental burning velocity** as the "burning velocity of a laminar flame under stated conditions of composition, temperature and pressure of the unburned gas."

Grossel (2002) provides considerably more detail on these definitions. He states:

> **Burning velocity** is the speed at which the flame front propagates relative to the unburned gas. This differs from flame speed. The **laminar burning velocity** is the speed at which a laminar (planar) combustion wave propagates relative to the unburned gas mixture ahead of it. The **fundamental burning velocity** is similar, but generally not identical to the observed laminar burning velocity. This is because (the fundamental burning velocity) is a characteristic parameter referring to standardized unburned gas conditions (normally 1 atm and 25°C), and which has been corrected for nonidealities in the measurement. The **turbulent burning velocity** exceeds the burning velocity measured under laminar conditions to a degree depending on the scale and intensity of turbulence in the unburned gas.

Clearly, care must be exercised in the usage of these terms.

2.1.1. Flammability Diagram

Hazard recognition for flammable materials involves determining the conditions (temperature, pressure and composition) under which the material is flammable.

For composition effects, a general way to represent the flammability of a gas or vapor is by a triangular diagram, as shown in Figure 2.7. Concentrations of fuel, oxygen and inert (in volume or mole %) are plotted on the three axis. Each apex of the triangle represents either 100% fuel, oxygen, or nitrogen. The tick marks on the scales show the direction in which the scale moves across the figure. Thus, point A represents a mixture composed of 60% methane, 20% oxygen and 20% nitrogen. The zone enclosed by the dashed line represents all mixtures which are flammable. Since point A lies outside the flammable zone, a mixture of this composition would not be flammable.

The air line on Figure 2.7 represents all possible combinations of fuel plus air. The air line intersects the nitrogen axis at 79% nitrogen (and 21% oxygen) which is the composition of pure air. The upper and lower flammability limits (in air) are shown as the intersection of the flammability zone boundary with the air line.

The stoichiometric line represents all stoichiometric combinations of fuel plus oxygen. The combustion reaction can be written in the form,

$$(1)\ \text{Fuel} + z\ O_2 \rightarrow \text{Combustion products} \tag{2-2}$$

where z is the stoichiometric coefficient for oxygen. The intersection of the stoichiometric line with the oxygen axis (in volume % oxygen) is given by

$$100\left(\frac{z}{1+z}\right) \tag{2-3}$$

The stoichiometric line is drawn from this point to the pure nitrogen apex.

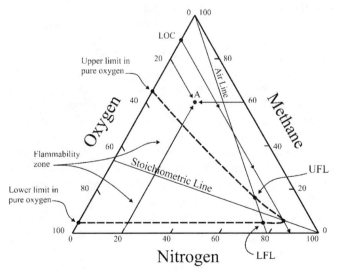

Figure 2.7. *Flammability diagram for methane at an initial temperature and pressure of 25°C and 1 atm (Mashuga and Crowl, 1998).*

Expression (2-3) is derived by realizing that, on the oxygen axis, no nitrogen is present. Thus, the moles present are fuel (1 mole) plus oxygen (z moles). The total moles are thus $1 + z$ and the mole or volume percent of oxygen is given by Expression (2-3) above.

An important point on Figure 2.7 is the **limiting oxygen concentration (LOC)**, shown as point LOC on the Figure. The LOC is determined by drawing a line just tangent to the nose of the flammability zone and parallel to the fuel axis. The LOC is the oxygen concentration below which fire or explosion is not possible for any mixtures. The LOC is frequently called the *minimum oxygen concentration (MOC)*, the *maximum oxygen concentration (MOC)*, the *maximum safe oxygen concentration (MSOC)*, or other names. The LOC depends on the fuel, temperature, pressure and inert species. Values of the LOC for a number of common materials are provided in Table 2.3. Also shown in the table are LOC values using carbon dioxide as the inert species—these values are different from the nitrogen values.

The shape and size of the flammability zone on a flammability diagram changes with a number of parameters, including the fuel, temperature, pressure and inert species. Thus, the flammability limits and LOC also change with these parameters.

Appendix A derives several equations that are useful for working with flammability diagrams. These results show that:

1. If two gas mixtures, R and S, are combined, the resulting mixture composition lies on a line connecting the points R and S on the flammability diagram. The location of the final mixture on the straight line depends on the relative moles in the mixtures combined—if mixture S has more moles, the final mixture point will lie closer to point S. This is identical to the lever rule, which is used for phase diagrams.
2. If a mixture R is continuously diluted with mixture S, the mixture composition will follow along the straight line between points R and S on the flammability diagram. As the dilution continues, the mixture composition will move closer and closer to point S. Eventually, at infinite dilution, the mixture composition will be at point S.
3. For systems having composition points that fall on a straight line passing through an apex corresponding to one pure component, the other two components are present in a fixed ratio along the entire line length.
4. The LOC can be estimated by reading the oxygen concentration at the intersection of the stoichiometric line and a horizontal line drawn through the LFL (see Appendix A). This is equivalent to the equation

$$LOC = z(LFL) \tag{2-4}$$

where z is the stoichiometric coefficient defined in Equation (2-2).

TABLE 2.3

Limiting Oxygen Concentrations (LOC)

These are the volume percent oxygen concentration above which combustion can occur.
(NFPA 69, 2002)

Gas or Vapor	N_2 /Air	CO_2 /Air
Methane	12	14.5
Ethane	11	13.5
Propane	11.5	14.5
n-Butane	12	14.5
Isobutane	12	15
n-Pentane	12	14.5
Isopentane	12	14.5
n-Hexane	12	14.5
n-Heptane	11.5	14.5
Ethylene	10	11.5
Propylene	11.5	14
1-Butene	11.5	14
Isobutylene	12	15
Butadiene	10.5	13
3-Methyl-1-butene	11.5	14
Benzene	11.4	14
Toluene	9.5	—
Styrene	9.0	—
Ethylbenzene	9.0	—
Vinyltoluene	9.0	—
Diethylbenzene	8.5	—
Cyclopropane	11.5	14
Gasoline		
(73/100)	12	15
(100/130)	12	15
(115/145)	12	14.5
Kerosene	10 (150°C)	13 (150°C)
JP-1 fuel	10.5 (150°C)	14 (150°C)
JP-3 fuel	12	14.5
JP-4 fuel	11.5	14.5
Natural gas	12	14.5
n-Butyl chloride	14	—
	12 (100°C)	—

(*continued on next page*)

TABLE 2.3 (*cont.*)

Gas or Vapor	N$_2$/Air	CO$_2$/Air
Methylene chloride	19 (30°C)	—
	17 (100°C)	—
Ethylene dichloride	13	—
	11.5 (100°C)	—
Methyl chloroform	14	—
Trichloroethylene	9 (100°C)	—
Acetone	11.5	14
n-Butanol	NA	16.5 (150°C)
Carbon disulfide	5	7.5
Carbon monoxide	5.5	5.5
Ethanol	10.5	13
2-Ethyl butanol	9.5 (150°C)	—
Ethyl ether	10.5	13
Hydrogen	5	5.2
Hydrogen sulfide	7.5	11.5
Isobutyl formate	12.5	15
Methanol	10	12
Methyl acetate	11	13.5
Methyl ether	10.5	13
Methyl formate	10	12.5
Methyl ethyl ketone	11	13.5
Vinyl chloride	13.4	—

Flammability diagrams are very useful to track the gas composition during a process operation to determine if a flammable mixture exists during the procedure. For example, consider a storage vessel containing pure methane whose inside walls must be inspected as part of its periodic maintenance procedure. For this operation, the methane must be removed from the vessel and replaced by air for the inspection workers to breathe. The first step in the procedure is to depressurize the vessel to atmospheric pressure. At this point the vessel contains 100% methane, represented by point A on Figure 2.8. If the vessel is now opened, and air allowed to enter, the composition of gas within the vessel will follow the air line on Figure 2.8 until the vessel gas composition eventually reaches point B, pure air. Note that at some point in this operation the gas composition passes through the flammability zone. If an ignition source of sufficient strength were present, then a fire or explosion would result.

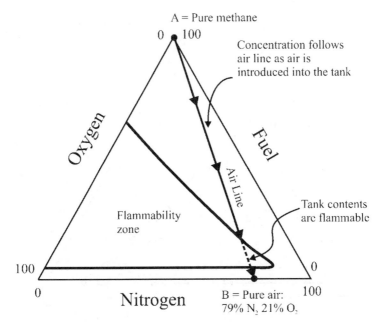

Figure 2.8. *The gas concentration during an operation to remove a vessel from service.*

The procedure is reversed for placing the vessel back into service. In this case the procedure begins at point B on Figure 2.8, with the vessel containing air. If the vessel is closed and methane pumped in, then the gas composition inside the vessel will follow the air line and end up at point A. Again the mixture is flammable as the gas composition moves through the flammability zone.

An inerting procedure can be used to avoid the flammability zone for both cases. This will be discussed in more detail in Section 3.4 on Inerting and Purging.

The determination of a complete flammability diagram requires several hundred tests using the device shown in Figure 2.3. Diagrams with experimental data for methane and ethylene are shown in Figures 2.9 and 2.10, respectively. Data in the center region of the flammability zone are not available since the maximum pressure exceeds the pressure rating of the vessel or unstable combustion or a transition to detonation is observed here. For these data, a mixture is considered flammable if the pressure increase after ignition is greater than 7% of the original ambient pressure, in accordance with ASTM E918 (1992). Note that many more data points are shown than required to define the flammability limits. This was done to obtain a more complete understanding of the pressure versus time behavior of the combustion over a wide range of mixtures. This information is important for mitigation of the explosion, should it occur.

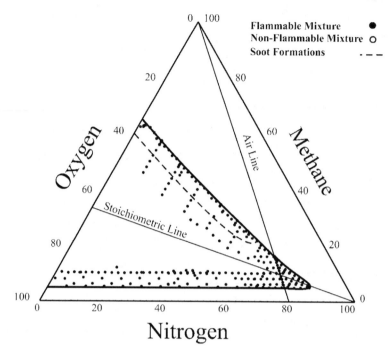

Experimental Conditions

Initial Pressure: 14.69 psia	Ignitor Type: 1cm 40 AWG SnCu / 500VA
Initial Temperature: 25°C	Ignitor Energy: 10 J
Reactor Volume: 20 liters	Ignitor Location: Center

Figure 2.9. *Experimental flammability diagram for methane (Mashuga and Crowl, 1999).*

A number of features are shown on Figures 2.9 and 2.10. First, the flammability zone is much larger for ethylene than methane—the upper flammability limit of ethylene is correspondingly higher. Second, the combustion produces copious amounts of soot in the upper, fuel-rich parts of the flammability zone. Finally, the lower boundary of the flammability zone is mostly horizontal and can be approximated by the LFL.

For most systems, detailed experimental data of the type shown in Figure 2.9 or 2.10 are unavailable. Several methods have been developed to approximate the flammability zone:

Method 1:

Shown by Figure 2.11

Given: Flammability limits in air, LOC, flammability limits in pure oxygen

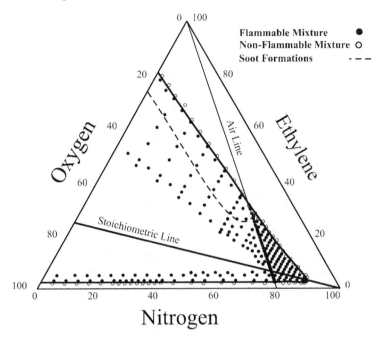

Experimental Conditions

Initial Pressure: 14.69 psia	Ignitor Type: 1cm 40 AWG SnCu / 500VA
Initial Temperature: 25°C	Ignitor Energy: 10 J
Reactor Volume: 20 liters	Ignitor Location: Center

Figure 2.10. *Experimental flammability diagram for ethylene (Mashuga and Crowl, 1999).*

Procedure:
1. Draw flammability limits in air as points on the air line.
2. Draw flammability limits in pure oxygen as points on oxygen scale.
3. Use Equation (2-3) to locate the stoichiometric point on the oxygen axis and draw the stoichiometric line from this point to the 100% nitrogen apex.
4. Locate the LOC concentration on the oxygen axis and draw a line parallel to the fuel axis until it intersects with the stoichiometric line. Draw a point at this intersection.
5. Connect all the points shown.

The flammability zone derived from this approach is only an approximation of the actual zone. Note that the lines defining the zone limits on Figures 2.9 and 2.10 are not exactly straight. This method also requires flammability limits in pure oxygen—data that might not be readily available. Flammability limits in pure oxygen for a number of common hydrocarbons are provided by Table 2.4.

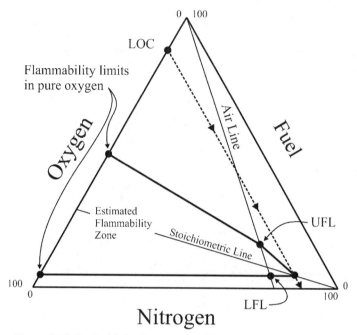

Figure 2.11. *Method 1 for the approximation of the flammability zone.*

TABLE 2.4
Flammability Limits in Pure Oxygen (Lewis and Von Elbe, 1987)

		Limits of Flammability in Pure Oxygen	
Compound	Formula	Lower	Upper
Hydrogen	H_2	4.0	94
Deuterium	D_2	5.0	95
Carbon monoxide[a]	CO	15.5	94
Ammonia	NH_3	15.0	79
Methane	CH_4	5.1	61
Ethane	C_2H_6	3.0	66
Ethylene	C_2H_4	3.0	80
Propylene	C_3H_6	2.1	53
Cyclopropane	C_3H_6	2.5	60
Diethyl ether	$C_4H_{10}O$	2.0	82
Divinyl ether	C_4H_6O	1.8	85

[a]The limits are insensitive to water vapor partial pressure above a few mm Hg.

Method 2:

Shown by Figure 2.12.

Given: Flammability limits in air, LOC

Procedure: Use steps 1, 3 and 4 from Method 1. In this case, only the points at the nose of the flammability zone can be connected. The flammability zone from the air line to the oxygen axis cannot be detailed without additional data, although it extends all the way to the oxygen axis and typically expands in size. The lower boundary can also be approximated by the LFL.

Method 3:

Shown by Figure 2.13.

Given: Flammability limits in air

Procedure: Use steps 1 and 3 from Method 1. Estimate the LOC using Equation (2-4). This is only an estimate, and usually (but not always) provides a conservative LOC.

For methods 2 and 3, only the flammability zone to the right of the air line can be drawn. As will be seen later, this is the portion of the flammability diagram that represents those situations most commonly encountered in preventing fires and explosions.

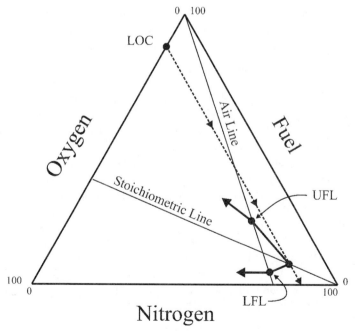

Figure 2.12. *Method 2 for the approximation of the flammability zone. Only the area to the right of the air line can be determined.*

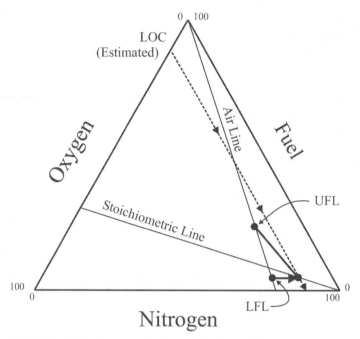

Figure 2.13. *Method 3 for the approximation of the flammability zone. Only the area to the right of the air line can be determined.*

2.1.2. Estimating Flammability Limits

Appendix C contains flammability limit data for a number of common materials. These limits were determined using a number of experimental methods. It is always recommended that the flammability limits be determined under conditions as close as possible to actual process conditions.

In many cases such data are difficult to obtain, or an immediate estimate is required for screening purposes. Several methods are available to estimate both the upper and lower flammability limits. These methods are satisfactory for most common hydrocarbons—their applicability to other materials has not been confirmed. Comparison of the predicted results with the actual data shows that the methods work better for the lower flammability limit than for the upper. The upper limit is typically more difficult to predict since more complex reaction chemistry occurs here.

Jones (1938) found that for many hydrocarbons, the upper and lower flammability limits are approximated by

$$LFL = 0.55C_{st} \tag{2-5}$$

$$UFL = 3.5C_{st} \tag{2-6}$$

where LFL and UFL are the lower and upper flammability limits, respectively (volume % fuel in air), and C_{st} is the stoichiometric concentration of fuel in air.

For the combustion reaction written by the stoichiometric equation,

$$C_mH_xO_y + z\,O_2 \rightarrow m\,CO_2 + (x/2)\,H_2O \qquad (2\text{-}7)$$

it follows that the stoichiometric coefficient, z, is given by,

$$z = m + x/4 - y/2 \qquad (2\text{-}8)$$

where z has units of moles O_2/mole fuel.

The stoichiometric percentage concentration in air is given by,

$$C_{st} = \frac{\text{moles fuel}}{\text{moles fuel} + \text{moles air}} \times 100$$

$$C_{st} = \frac{100}{1 + \left(\dfrac{\text{moles air}}{\text{moles fuel}}\right)}$$

$$= \frac{100}{1 + \left(\dfrac{1\ \text{mole air}}{0.21\ \text{mole }O_2}\right)\left(\dfrac{\text{moles }O_2}{\text{moles fuel}}\right)}$$

$$C_{st} = \frac{100}{1 + (z/0.21)} \qquad (2\text{-}9)$$

where the number 0.21 represents the oxygen content in air.

Substituting z from Equation (2-8) into Equation (2-9), and substituting the result into Equations (2-5) and (2-6), results in the following equations for estimating the flammability limits.

$$LFL = \frac{0.55(100)}{4.76m + 1.19x - 2.38y + 1} \qquad (2\text{-}10)$$

$$UFL = \frac{3.5(100)}{4.76m + 1.19x - 2.38y + 1} \qquad (2\text{-}11)$$

Another method (Suzuki, 1994; Suzuki and Koide, 1994) correlated the flammability limits as a function of the heat of combustion of the fuel. A good fit was obtained for 112 organic materials containing carbon, hydrogen, oxygen, nitrogen and sulfur. The resulting correlations are,

$$LFL = \frac{-3.42}{\Delta H_c} + 0.569\Delta H_c + 0.0538\Delta H_c^2 + 1.80 \qquad (2\text{-}12)$$

$$UFL = 6.30\Delta H_c + 0.567\Delta H_c^2 + 23.5 \qquad (2\text{-}13)$$

where LFL and UFL are the lower and upper flammability limits, respectively (volume % fuel in air), and ΔH_c is the heat of combustion for the fuel, in 10^3 kJ/mol.

Equation (2-13) is applicable only over the UFL range of 4.9% to 23%. If the heat of combustion is provided in kcal/mole, it can be converted to kJ/mol by multiplying by 4.184.

The prediction capability of Equations (2-5) through (2-13) is only modest, at best. For hydrogen the predictions are poor. For methane and the higher hydrocarbons the results are improved. Thus, these methods should only be used for a quick initial estimate, and should not replace actual experimental data.

Flammability limits can also be estimated using calculated adiabatic flame temperatures (CAFT) (Hansel, Mitchell et al., 1991; Melhem, 1997). The procedure for calculating the adiabatic flame temperature is described in Section 2.6 on Kinetics and Thermochemistry. It is based on the premise that the flammability limits are mostly thermal in behavior and are not highly dependent on kinetics. A typical adiabatic flame temperature of 1200K is used to define the flammability limits, although other values are used depending on whether more or less conservative results are required. The CAFT approach is capable of estimating the flammability behavior for any mixture of gases, including air. Thus, it is possible to estimate the entire flammability zone in a triangle diagram using this approach. Unfortunately, this has not been extensively studied as of this writing.

Britton (2002b) proposed using a heat of oxidation to estimate flammability limits. The heat of oxidation is defined as the heat of combustion divided by the stoichiometric ratio for oxygen. Britton found that the heat of oxidation can be correlated with the flammability limits for hydrocarbons.

2.1.3. Temperature Effect on Flammability

In general, the size of the flammability zone shown in Figure 2.7 increases with increasing temperature. For mixtures in air, the UFL increases and the LFL decreases, broadening the range over which the mixture is flammable. Some materials (such as decane) which are not flammable at ambient conditions may become flammable at increased temperature.

A set of empirical equations to estimate the effect on the flammabiltiy limits with temperature are available (Zabetakis, Lambiris et al., 1959)

$$\text{LFL}_T = \text{LFL}_{25} - \frac{0.75}{\Delta H_c}(T - 25) \qquad (2\text{-}14)$$

$$\text{UFL}_T = \text{UFL}_{25} + \frac{0.75}{\Delta H_c}(T - 25) \qquad (2\text{-}15)$$

where LFL_T is the lower flammability limit at temperature T (volume % fuel in air)

LFL_{25} is the lower flammability limit at 25°C (volume % fuel in air)

ΔH_c is the heat of combustion of the fuel (kcal/mole)

T is the temperature (°C)

UFL_T is the upper flammability limit at temperature T (volume % fuel in air)

UFL_{25} is the upper flammability limit at 25°C (volume % fuel in air).

2.1.4. Pressure Effect on Flammability

In general, pressure has little effect on the LFL, except at low pressures (below about 50 mm Hg for most vapors) where combustion is not possible. The UFL generally increases as the pressure increases and decreases as the pressure decreases. As the pressure increases the flammability range generally increases. Some materials (such as jet fuel) which are not flammable at ambient pressure may become flammable at increased pressure.

An empirical expression to estimate the change in upper flammability limit with pressure is available (Zabetakis, 1965)

$$UFL_P = UFL + 20.6(\log P + 1) \tag{2-16}$$

where UFL_P is the upper flammability limit at pressure P (volume % fuel in air), UFL is the upper flammability limit at 1 atm (volume % fuel in air), and P is the pressure (mega Pascals absolute).

2.1.5. Flammability of Gaseous Mixtures

The flammable behavior of mixtures of gases is not completely understood at this time—it is best to determine this behavior experimentally under conditions as close as possible to process conditions.

There are two methods commonly used to estimate the flammability limits of mixtures.

The first method is called Le Chatelier's rule (Le Chatelier, 1891). The following empirical equations for the flammability limits are provided.

$$LFL_{mix} = \left(\sum_{i=1}^{n} \frac{y_i}{LFL_i} \right)^{-1} \tag{2-17}$$

$$UFL_{mix} = \left(\sum_{i=1}^{n} \frac{y_i}{UFL_i} \right)^{-1} \tag{2-18}$$

where LFL_{mix} is the lower flammability of the mixture (volume % fuel in air)
   LFL_i is the lower flammability limit for flammable species i (volume % fuel in air)
   UFL_{mix} is the upper flammability limit of the mixture (volume % fuel in air)
   UFL_i is the upper flammability limit for flammable species i (volume % fuel in air)
   y_i is the mole fraction for flammable species i on a flammable species basis

Le Chatelier's rule provides flammability limit estimates that are close to experimental values for many simple hydrocarbons, but has no fundamental basis. The method applies only to mixtures with air.

The second approach uses the calculated adiabatic flame temperature (CAFT) method discussed in Section 2.1.2 and in more detail in Section 2.6 on Kinetics and Thermochemistry. In this case, a commercial equilibrium software package is recommended to perform the extensive calculations since the number of species involved is typically large.

2.1.6. Minimum Ignition Energies

The **minimum ignition energy (MIE)** is defined as the "minimum amount of thermal energy released at a point in a combustible mixture that will cause indefinite flame propagation away from that point, under specified test conditions" (NFPA 68, 1998). The ignition of a flammable material is a complex subject which is discussed in more detail elsewhere (Lewis and von Elbe, 1987; Glassman, 1996). Using a simplistic analysis, adequate energy must be provided to a small volume of flammable mass over a short enough period of time to increase the temperature to the point where the reaction generates enough energy to sustain itself. Since complex chain branching reactions are possible, a time delay or **induction time** might be exhibited.

The MIE depends on the specific chemical or chemicals, concentrations, pressure, temperature and the mode of ignition. For a spark ignition, the MIE depends also on the spark gap size and the duration of the spark. Thus, from a practical standpoint, the MIE is difficult to characterize and is highly dependent on the experimental configuration. Limited MIE data are available and considerable variation between researchers is noted. Table 2.5 provides MIE data for a number of common combustible gases. More information is provided elsewhere (Lewis and von Elbe, 1987; Glassman, 1996).

In general, the following statements apply to practical usage,

- For most flammable gases, the MIE is typically no lower than 0.10 mJ, although 0.25 mJ is commonly used. Specific chemicals (e.g., hydrogen) have lower values.

TABLE 2.5
Minimum Ignition Energy (MIE) for Selected Gases (Glassman, 1996)

Chemical	Minimum Ignition Energy (mJ)
Acetylene	0.020
Benzene	0.225
1,3-Butadiene	0.125
n-Butane	0.260
Cyclohexane	0.223
Cyclopropane	0.180
Ethane	0.240
Ethene	0.124
Ethylacetate	0.480
Ethylene oxide	0.062
n-Heptane	0.240
Hexane	0.248
Hydrogen	0.018
Methane	0.280
Methanol	0.140
Methyl acetylene	0.120
Methyl ethyl ketone	0.280
n-Pentane	0.220
2-Pentane	0.180
Propane	0.250

- For flammable gases in air, the lowest MIE (LMIE) is found near the stoichiometric concentration (but not necessarily at the stoichiometric concentration) and the MIE increases as the fuel concentration increases or decreases from the LMIE value (Britton, 1999).
- For most flammable dusts, a typical value for the MIE is about 10 mJ, although wide variability is expected depending on dust type, particle size, etc.
- As the temperature increases, the MIE decreases.
- As the pressure increases, the MIE decreases.
- An increase in inert gas concentration increases the MIE.

The most important concept to remember is that the energy required to ignite a flammable gas or vapor is very low. The MIE of 0.25 mJ represents the kinetic energy contained in a small coin as it impacts a surface after being dropped from a height of a few millimeters A static electricity discharge felt by a person has an energy of greater than about 20 mJ—orders of magnitude greater than the MIE. Thus, flammable gases and vapors are readily ignited. This is the principle reason why the elimination of ignition sources cannot be relied on as the primary defense against fires and explosions of gases and vapors.

2.1.7. Autoignition Temperature

As the temperature of a flammable mixture of gas or vapor is increased, a temperature is eventually reached where the mixture will ignite without the need for an external ignition source—the mixture will ignite spontaneously. This temperature is called the **autoignition temperature (AIT)**. A fuel at a high enough temperature, when released and mixed with air to form a flammable mixture, can ignite in this fashion. A cold fuel released and mixed with hot air or contacting a hot surface will also autoignite.

The AIT depends on many factors, including the particular fuel, pressure, test vessel volume, presence of catalytic materials, flow conditions and so forth. It is important to determine the AIT under test conditions that replicate as close as possible the actual process conditions.

Appendix C provides AIT values for a number of chemicals. Wide variability exists in published AIT values—care must be employed in their use. If a number of differing literature values are reported, the lowest value is usually selected to ensure a conservative design.

In general, the AIT

- Decreases with increasing pressure.
- Increases as mixtures become rich or lean.
- Decreases with increased oxygen concentration.
- Decreases in the presence of catalytic surfaces (which may include rusty pipes) and in the presence of combustion sensitizers (like NO_x).

The AIT also typically decreases as the test volume increases and is lower under flow conditions. An ignition delay or induction time is also possible.

2.1.8. Example Applications

EXAMPLE 2.1

A vessel contains a gas mixture composed of 50% methane and 50% nitrogen. If the mixture escapes from the vessel and mixes with air, will it become flammable?

Solution A flammability diagram for this case is shown in Figure 2.14. The initial concentration of the mixture is denoted as point A. As the gas escapes and mixes with air the composition will follow the straight line shown, eventually

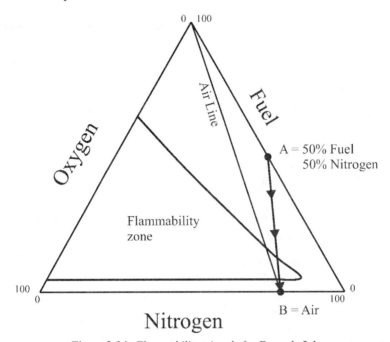

Figure 2.14. *Flammability triangle for Example 2.1.*

diluting to the point of becoming pure air—shown as point B on Figure 2.14. Since the straight line moves through the flammability zone, the gas mixture will at some point become flammable. Thus, if an ignition source of sufficient energy is present, a fire or explosion may result.

EXAMPLE 2.2
Estimate the LFL and UFL for hydrogen, methane, butane, and hexane and compare to published values. Use two methods: Equation (2-10) and (2-11) and Equation (2-12) and (2-13).
Solution: The two methods are applied using the equations indicated. The complete results are shown in Table 2.6. The experimental values for the flammability limits were obtained from Appendix C. As expected, the predictive methods produce only modest results, when compared to actual experimental data.

EXAMPLE 2.3
Estimate the UFL and LFL for a vapor mixture comprised of 63.5% ethyl acetate, 20.8% ethyl alcohol and 15.7% toluene, by volume.
Solution: The flammability limits for these species are provided in Appendix C. These are shown in the following table:

Species	LFL (Vol. % fuel in air)	UFL (Vol. % fuel in air)
Ethyl Acetate, $C_4H_8O_2$	2.5	9
Ethyl Alcohol, C_2H_6O	4.3	19
Toluene, C_7H_8	1.4	6.7

Equations (2-17) and (2-18) are used to determine the flammability limits of the mixture,

$$LFL_{mix} = \left(\sum_{i=1}^{3} \frac{y_i}{LFL_i} \right)^{-1} = \left(\frac{0.635}{2.5} + \frac{0.208}{4.3} + \frac{0.157}{1.4} \right)^{-1} = 2.41\%$$

$$UFL_{mix} = \left(\sum_{i=1}^{3} \frac{y_i}{UFL_i} \right)^{-1} = \left(\frac{0.635}{9.0} + \frac{0.208}{19.0} + \frac{0.157}{6.7} \right)^{-1} = 9.5\%$$

The experimentally determined value of the LFL is 2.04% (Lewis and von Elbe, 1987). Thus, the prediction is only fair.

EXAMPLE 2.4 (CROWL AND LOUVAR, 1990)
A gas mixture is composed of 2.0% methane, 0.8% hexane and 0.5% ethylene, by volume. The remaining gas is air. Is the mixture flammable?

TABLE 2.6
Results for Example 2.2

Species	Stoichi- ometry	Heat of Combustion 10^3 kJ/mol	Experimental % Fuel in air		Eqs. (2-10), (2-11)		Eqs. (2-12), (2-13)	
			LFL	UFL	LFL	UFL	LFL	UFL
Hydrogen, H_2	$m = 0$ $x = 2$ $y = 0$	−0.2418	4	75	16.3	103	15.8	22.0
Methane, CH_4	$m = 1$ $x = 4$ $y = 0$	−0.8903	5.3	15.0	5.2	33.3	5.2	18.3
Butane, C_4H_{10}	$m = 4$ $x = 10$ $y = 0$	−2.877	1.9	8.5	1.6	8.4	1.8	10.1
Hexane, C_6H_{14}	$m = 6$ $x = 14$ $y = 0$	−4.1945	1.1	7.5	1.2	7.69	1.2	7.0

Solution: Equations (2-17) and (2-18) are used to estimate the flammability limits of the mixture. These equations require the mole fraction *on a flammable basis only.* This is determined as follows:

Species	Volume %	Mole Fraction on Combustible Basis
Methane, CH_4	2.0	2.0/3.3 = 0.61
Hexane, C_6H_{14}	0.8	0.8/3.3 = 0.24
Ethylene, C_2H_4	0.5	0.5/3.3 = 0.15
Total Combustibles	3.3	

The LFL and UFL for each combustible species are obtained from Appendix C. The required values are shown in the table below,

Species	LFL	UFL
Methane	5.3	15
Hexane	1.2	7.5
Ethylene	3.1	32

Equation (2-17) is used to estimate the LFL,

$$\text{LFL}_{\text{mix}} = \left(\sum_{i=1}^{3} \frac{y_i}{\text{LFL}_i} \right)^{-1} = \left(\frac{0.61}{5.3} + \frac{0.24}{1.2} + \frac{0.15}{3.1} \right)^{-1} = 2.8\%$$

Equation (2-18) is used to estimate the UFL,

$$\text{UFL}_{\text{mix}} = \left(\sum_{i=1}^{3} \frac{y_i}{\text{UFL}_i} \right)^{-1} = \left(\frac{0.61}{15} + \frac{0.24}{7.5} + \frac{0.15}{32.0} \right)^{-1} = 12.9\%$$

The total combustibles present are 2.0 + 0.8 + 0.5 = 3.3%, which is between the flammability limits. Thus, we can expect the mixture to be flammable.

2.2. Liquids

An important flammability characteristic for liquids is the **flashpoint temperature**, or simply the **flashpoint**. The flashpoint temperature is useful for determining the flammability hazard of the liquid.

The flashpoint temperature could be experimentally determined as follows: A flammable liquid is placed in an open cup, as shown in Figure 2.15. The liquid and container are slowly heated by a Bunsen burner below the cup. A small flame on

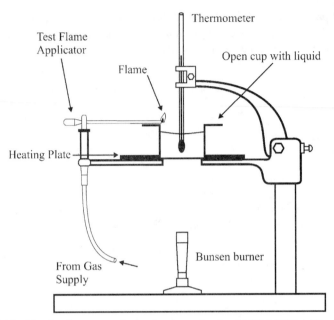

Figure 2.15. *Cleveland open cup flashpoint determination (ASTM 1990a). The test flame applicator is moved back and forth horizontally over the liquid sample.*

the end of a wand provides an ignition source immediately above the liquid surface. Initially, when the wand is moved across the cup no flame or flash is produced. As the temperature of the liquid increases, the vapor pressure increases. Eventually a temperature is reached where the liquid produces enough vapor for the air/vapor mixture above the liquid level to become flammable. This temperature is the flashpoint temperature. At this temperature, the liquid is generally not capable of producing adequate vapor to support a continuous flame on its surface—a flash is typically observed and the flame goes out. As the temperature is increased further, the **fire point** temperature is reached where a continuous flame or combustion is supported.

Appendix C contains flashpoint temperatures for a number of common materials. In general, the lower the flashpoint temperature, the more hazardous the material. If a liquid is at a temperature above its flashpoint temperature, then sufficient vapor is present to form a flammable mixture with air. If a liquid with a flashpoint temperature below room temperature is stored in an open container, then flammable vapors are likely to exist above the liquid—only an ignition source is required to result in a fire or explosion.

The apparatus shown in Figure 2.15 is one of the simplest flashpoint determination methods (ASTM D9290, 1990a). It is called an open cup method for obvious reasons. Open cup methods suffer from air drafts above the cup that reduce the vapor concentration. Several other experimental methods are available to determine flashpoint temperatures that reduce the draft effect (ASTM D5687, 1987; ASTM D3278, 1989; ASTM D9390, 1990b; ASTM E918, 1992). These methods involve containers that are initially closed, with a small shutter which opens to expose the vapor to the ignition source. Because these methods reduce the draft, the flashpoint temperature measured is generally lower than with open cup methods. The procedure used with many of these methods is trial and error. A temperature is set on the equipment and if a flash is not observed when the shutter is opened the test is repeated at a higher temperature with new liquid material. An accurate determination of the flashpoint using these methods requires many tests producing a modest quantity of liquid waste.

The determination of the flashpoint using a standard closed cup method can be completed in several hours using relatively low cost equipment. The chemical species or mixture tested must match closely the species or mixture actually present in the process.

The standard flashpoint is always calibrated and listed at 1 atm total pressure. It is possible to determine a flashpoint at other pressures. In general, as the pressure decreases, the flashpoint temperature decreases and vice versa—at the same temperature and lower pressure more of the liquid will vaporize eventually providing a flammable concentration. The flashpoints for materials processed in a facility at increased elevation would be lower than for the same materials processed at sea level. Thus, diesel fuel, which is normally not flammable at sea level, may become flammable as the truck drives into high altitudes.

Figure 2.16 is a plot of fuel vapor concentration versus temperature and shows how the flashpoint temperature is related to the vapor pressure curve and the upper and lower flammability limits. The saturation vapor pressure curve is exponential and increases rapidly as the temperature increases. The lower flammability limit theoretically intersects the vapor pressure curve at the flashpoint temperature. At higher temperatures an autoignition region is found—the autoignition temperature (AIT) is the lowest temperature in this region.

From a practical standpoint, the flashpoint temperatures reported frequently do not correspond exactly with the intersection of the lower flammability limit with the vapor pressure curve. This is due to the experimental methods used to determine flashpoints. In general, the closed cup flashpoint values are in closer agreement with the lower flammability limit.

NFPA 30 (2000) uses the flashpoint temperature to define the terms **combustible** and **flammable**. A combustible liquid is any liquid having a flashpoint temperature at or above 100°F (37.8°C). A flammable liquid is one with a flashpoint below 100°F. Thus, kerosene (with a flashpoint of 104°F) would be classified as combustible, while gasoline (with a flashpoint well below room temperature) is classified as flammable.

The U. S. Department of Transportation (DOT) defines a liquid as flammable if its flashpoint is not more than 141°F (60.5°C) (DOT, 2000). In practice, the head

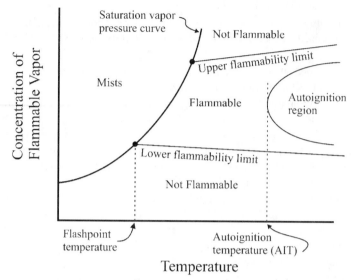

Figure 2.16. *The relationship between the various flammability properties.*

space of storage tanks subject to solar radiation can easily reach 100°F and, consequently, the DOT value is perhaps more realistic than NFPA's 100°F and is certainly not overly conservative.

Many substances do not have flashpoints and are considered not flammable under NFPA 30 and the DOT rules. However, occasionally someone attempts to weld or cut into a vessel which contains such a material and ignites the vapors with disastrous results. Under certain circumstances, if a material does not have a flashpoint, it doesn't always mean that the vapors can't be ignited under all conditions.

If experimental flashpoint data are not available, Satyanarayana and Rao (1992) provide an empirical method for different groups of organic compounds. They tested their correlation with 1221 compounds and found less than 1% absolute error with experimental data (based on °K).

2.2.1. Flashpoints of Mixtures of Liquids

For miscible liquid mixtures comprised of a single flammable liquid mixed with nonflammable liquids, the flashpoint of the mixture will occur at a temperature where the partial pressure of the flammable liquid in the mixture is equal to the vapor pressure of the pure liquid at its flashpoint temperature.

The procedure to estimate a flashpoint temperature for this type of mixture is as follows:

1. Given: mole fraction of flammable liquid species i in the liquid phase, x_i; flashpoint temperature for pure liquid, T_F; total pressure.

2. Calculate the saturation vapor pressure of the pure flammable liquid at its flashpoint temperature, $P_i^{sat}(T_F)$, using published vapor pressure data or equations.

3. Guess a flashpoint temperature, T_G, of the mixture.

4. Calculate the saturation vapor pressure of the pure flammable liquid at the guessed temperature, $P_i^{sat}(T_G)$, using published vapor pressure equations.

5. Use a vapor–liquid equilibrium model to estimate the partial pressure of the flammable vapor above the liquid mixture, p_i. If the liquid mixture is ideal, this is given by,

$$p_i = x_i P_i^{sat}(T_G) \tag{2-19}$$

where p_i is the partial pressure of the flammable vapor, x_i is the mole fraction of the flammable liquid in the mixture, and $P_i^{sat}(T_G)$ is the saturation vapor pressure of the flammable liquid at the guessed temperature.

6. If the partial pressure, p_i, is less than $P_i^{sat}(T_F)$, then the guessed temperature is too low. If the reverse is true, then the guessed temperature is too high.

7. Guess a new value for the flashpoint temperature and return to step 4. Repeat the procedure until a temperature of suitable precision is determined.

For many liquids the saturation vapor pressure curve is given by the Antoine equation,

$$\ln(P_i^{sat}) = A - \frac{B}{C+T} \tag{2-20}$$

where A, B, and C are constants and T is the absolute temperature.

Then, for ideal systems represented by Equation (2-19), the flashpoint condition is given by,

$$p_i = x_i P_i^{sat}(T_G) = P_i^{sat}(T_F) \tag{2-21}$$

Equation (2-21) is combined with Equation (2-20) and the result is either numerically or algebraically solved for T_G.

For mixtures composed of several flammable liquids, steps 5 and 6 are replaced as follows:

5. Use a vapor–liquid equilibrium procedure to estimate the partial pressures of all the flammables in the vapor phase. Convert the partial pressures to mole fractions and determine the mole fraction for each flammable species on a combustible basis only.

6. Use Le Chatelier's law, Equations (2-17) and (2-18), to determine the flammability of the vapor mixture. If the mixture composition is below the LFL for the mixture, the guessed temperature is too low. If the reverse is true, then the guessed temperature is too high.

The application of the above procedure to complex mixtures has not been extensively tested. From a practical standpoint, the flashpoint procedure is simple enough that an experimental determination is almost always done.

A common industrial practice is to assume that the flashpoint of the mixture is never less than the lowest flashpoint of any of the pure components. For ideal systems and most nonideal systems, this is true. However, for minimum boiling point azeotropes, the mixture flashpoint can be significantly less than any of the pure component flashpoints (Larson, 1998).

A more detailed method for estimating flashpoints of nonideal liquid mixtures is provided by Henley (1998).

For nonmiscible liquids the flashpoint may be dictated by the phase structure of the mixture and is a complex problem beyond the scope of this book.

2.2.2. Example Applications

EXAMPLE 2.5
Determine the flashpoint temperature of a liquid mixture of 50 mole percent methanol and water.

Solution: The flashpoint for pure methanol is given in Appendix C as 12.8°C. The vapor pressure for methanol is given in Antoine form, Equation (2-20), by Crowl and Louvar (2002)

$$\ln(P_i^{\text{sat}}) = 18.5875 - \frac{3626.55}{-34.29 + T}$$

where P_i^{sat} is the saturation vapor pressure, in mm Hg and T is the temperature in K.

At the flashpoint temperature of 12.8°C = 286 K, the vapor pressure of the pure methanol is

$$\ln(P_i^{\text{sat}}) = 18.5875 - \frac{3626.55}{-34.29 + 286} = 4.18$$

$$P_i^{\text{sat}} = e^{4.18} = 65.4 \text{ mm Hg}$$

The flashpoint of the mixture will occur at the temperature at which the partial pressure of the methanol above the mixture is equal to 65.4 mm Hg. From Equation (2-21), assuming an ideal liquid mixture,

$$(0.50)P_i^{\text{sat}}(T_G) = 65.4 \text{ mmHg}$$

$$P_i^{\text{sat}}(T_G) = 65.4/0.50 = 131 \text{ mm Hg}$$

Equation (2-20) is used to determine the new flashpoint temperature,

$$\ln(131) = 18.5875 - \frac{3626.55}{-34.29 + T}$$

Solving for T results in a flashpoint of the mixture of 299 K or 26°C. The flashpoint of the mixture increases as water is added.

This example assumes ideal vapor liquid behavior which is not valid for all concentration ranges for this system. More complex vapor–liquid equilibrium methods must be used with the same approach for nonideal systems.

2.3. Aerosols and Mists

Aerosols are liquid droplets or solid particles of size small enough to remain suspended in air for prolonged periods of time. **Mists** are suspended liquid droplets produced by condensation of vapor into liquid or by the breaking up of a liquid into a dispersed state by splashing, spraying or atomizing (Olishifski, 1971). Aerosols or mists are frequently formed when liquids are discharged from process equipment under high pressure, or when pressurized liquified gases, such as LPG, flash into vapor when the pressure is suddenly reduced.

The combustion behavior of aerosols and mists is not well understood. Combustion models representing the behavior of a small, single drop of liquid match experimental data quite well. Efforts to model the combustion behavior of clouds of droplets have not been as successful due to droplet interactions and the turbulence created by the rapid generation and expansion of the combustion gases— phenomena which are not well understood.

It is known that liquids in aerosol or mist form may burn or explode at temperatures well below their flashpoint temperature—the behavior shown in Figure 2.16 does not apply to aerosols and mists. The common home oil furnace uses the atomization of room temperature oil to produce a combustible mist. The concepts of vapor flammability limits and liquid flashpoint temperatures are not adequate to describe the flammability of aerosols or mists. NFPA 68 (2002) uses the suggestion that a typical LFL for a fine hydrocarbon mist is 40 to 50 gm/m^3, which is approximately equal to the LFL for combustible hydrocarbon gases in air at room temperature.

2.4. Dusts

A **dust** is defined as "any finely divided solid, 420 μm or 0.016 in, or less in diameter (that is, material capable of passing through a U.S. No. 40 standard sieve) (NFPA 68, 2002)

The dusts of most solid combustible materials, when dispersed in air and ignited, can burn rapidly, introducing the potential for a dust explosion. Wood, aluminum, coal, flour, milk powder, dyes and pharmaceuticals are examples of materials that have caused explosions. Grain dust is combustible and is responsible for explosions in grain elevators. Many chemicals used and produced by the chemical industry are in dust form. The explosive behavior of these materials should be adequately characterized prior to processing, handling, shipping and use.

Dusts can present both fire and explosion hazards. First, dust layers, which collect on hot equipment surfaces may heat, smoulder and catch on fire. The dust layer provides thermal insulation to the equipment resulting in higher surface temperatures and increased likelihood for ignition. Second, dust particles may entrain with air forming a combustible dust cloud.

Dusts are unique in that other properties affect their combustion behavior, including moisture content, particle size, particle size distribution, particle loading, etc. Sometimes these additional properties are combined with the fire triangle of Figure 2.2 to form a fire pentagon or hexagon for dusts. These additional properties make the characterization of dusts very difficult—experimental characterization at conditions as close as possible to process conditions is highly recommended.

It is commonly believed that dusts burn in a fashion similar to gases—a belief that is usually not correct. For instance, suppose a material combusts within a large chamber and the combustion is relieved through a vent in the side of the chamber. If the combustible material is a flammable gas, the fireball exiting the vent will be relatively small and can be almost transparent. If the material is a dust, most of the dust will burn outside the chamber, resulting in a much larger, opaque flame and considerable smoke. The dust burns external to the chamber due to two phenomena. First, the dust contains a much larger quantity of fuel than the gas and the oxygen content within the chamber is typically inadequate to combust all of the dust. The combustion is completed by the additional oxygen available once the unburnt dust is ejected from the chamber. Second, most dusts burn slower than gas, with much of the dust burning after it is ejected from the chamber.

Figure 2.17 shows an apparatus used to characterize the pressure–time behavior of dust cloud explosions. It is similar to the apparatus in Figure 2.3 used for vapors. A minimum test vessel volume of 20 liters is recommended in order to minimize the effects of surface quenching and to allow for scale-up of the results. The dust is initially held in a container outside the vessel and blown into the vessel by air. The initial pressure of the vessel is selected so that the addition of the dust blowing gas results in a total absolute pressure of 1 atm just prior to ignition. The initial turbulence of the mixture can have a large effect on the experimental results, but the dust must also be properly dispersed and mixed with air prior to ignition. Thus, a time delay is set between the time the dust is blown in and the time the igniter is activated so that much of the turbulence is dissipated before ignition. However, the time delay must not be so long that the dust begins to settle out. Once the igniter is activated the pressure history of the explosion is tracked by a pressure transducer.

Figure 2.18 shows a typical pressure–time data curve from the dust cloud explosion apparatus. The pressure rises rapidly to a maximum over a period of

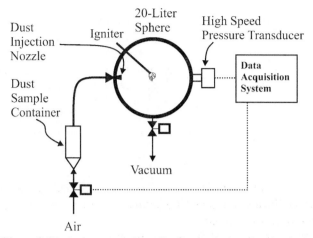

Figure 2.17. *An apparatus for collecting explosion data for dusts.*

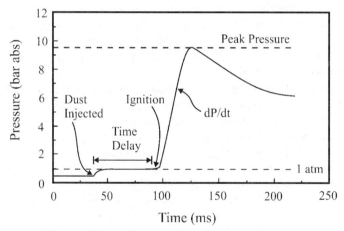

Figure 2.18. *Pressure data from dust explosion device.*

milliseconds and then decays slowly as the product gases are cooled by the vessel surface. The behavior is very similar to the pressure history for the vapor explosion apparatus shown in Figure 2.4.

A deflagration index is defined for dusts in an identical fashion to vapors. For dusts, the deflagration index is denoted by K_{St} where the "St" stands for "Staub," the German word for dust.

$$K_{St} = \left(\frac{dP}{dt} \right)_{max} V^{1/3} \qquad (2\text{-}22)$$

As the K_{St} value increases, the dust explosion becomes more violent.

TABLE 2.7
St-Classes for Dusts

Deflagration Index K_{St} (bar-m/sec)	St-class	Examples
0	St-0	Rock dust
1-200	St-1	Wheat grain dust
200-300	St-2	Organic dyes
>300	St-3	Aspirin, aluminum powder

Table 2.7 shows how the K_{St} values are organized into four St-classes. The St-class number increases as the deflagration index increases, that is, as the dust explosion becomes more violent.

Appendix E contains combustion data for a number of dust materials. The test vessel volume, moisture content, igniter energy, particle shape and particle surface reactivity of the materials in Appendix E are unknown and could result in wide variability in these results. Thus, the applicability of the data to apparently identical materials is limited. Appendix E includes the median particle size of the dust tested and the dust concentration under test conditions—important parameters in dust explosion characterization.

In general, as the particle size or moisture content decreases, the deflagration index, K_{St}, and the maximum pressure, P_{max}, increase while the minimum explosible dust concentration (MEC), C_{min}, and minimum ignition energy (MIE), decrease. Over a limited range of particle size, reducing the particle size has more effect on K_{St} and MIE than on P_{max}. As the initial pressure increases, the maximum pressure and, under certain conditions, the maximum rate of pressure rise, generally increase proportionally. The minimum ignition energy (MIE) generally decreases with increased initial temperature. As the oxygen concentration decreases, the MIE increases, the maximum pressure and maximum rate of pressure rise of the dust in general decrease, with the reverse occurring for oxygen enriched systems (Eckhoff, 1997).

Dust clouds also exhibit a limiting oxygen concentration (LOC) below which combustion of the dust cloud is not possible. LOC values for a number of dusts are shown in Table 2.8.

Dust explosions in mines, grain elevators and other processing facilities can occur in two or more stages. The first stage involves an explosion, which may be small and which may or may not be related to the dust. This first stage of the explosion suspends dust into the air which may be laying undisturbed within the process or in surrounding areas. The dust mixes with air and forms a combustible mixture. The second stage of the explosion then involves the suspended dust, which is frequently larger and more damaging than the first stage explosion. Additional stages of the explosion may occur as more dust is disturbed.

TABLE 2.8
*Limiting Oxygen Concentration (LOC – vol. %) for Dust Clouds
in Oxygen/Nitrogen Atmospheres (Eckhoff, 1997)*

Dust Type	Median Particle Size (μm)	Minimum O_2 Concentration
Cellulosic Materials		
Cellulose	51	11
Wood	27	10
Food and Feed		
Maize (corn) starch	17	9
Rye flour 1150	29	13
Coals		
Brown coal	42	12
Brown coal	63	12
Brown coal	66	12
Bituminous coal	17	14
Plastics, Resins, Rubber		
Polyacrylnitrile	26	11
Polyacrylnitrile	26	10
Pharmaceuticals, Pesticides		
Methionine	<10	12
Intermediate Products, Additives		
Barium stearate	<63	13
Benzoyl peroxide	59	10
Methyl cellulose	29	15
Methyl cellulose	49	14
Methyl cellulose	70	10
Paraformaldehyde	23	6
Other Technical Chemical Products		
Blue dye	<10	13
Organic pigment	<10	12
Metals, Alloys		
Aluminum	22	5
Aluminum	22	6
Magnesium alloy	21	3
Other Inorganic Products		
Soot	<10	12
Soot	13	12
Soot	16	12

There are many mechanisms for the formation of dusts. Many solids are friable with dust formation during processing and handling. This includes mechanical transfer during loading and unloading, pneumatic transport, sieving, blending and dumping. Segregation and particle size classification can also occur during pneumatic transfer—clouds enriched with finer particles can be created resulting in an explosion hazard not indicated by the bulk material. For these reasons, dust explosion testing is often performed on material of much smaller particle size to give a conservative result.

Clearly, good housekeeping in processing areas is important to prevent and mitigate dust explosions. Special procedures and equipment are required to transfer combustible dusts into and out of processing equipment and to process these materials safely (NFPA 654, 2000).

Several books discuss dust explosions (Bartknecht, 1981; Eckhoff, 1997) and numerous NFPA publications are available (NFPA 482, 1996; NFPA 499, 1997; NFPA 480, 1998; NFPA 651, 1998; NFPA 664, 1998; NFPA 61, 1999; NFPA 120, 1999; NFPA 481, 2000; NFPA 654, 2000; NFPA 655, 2001).

2.5. Hybrid Mixtures

NFPA 68 (1994) defines a **hybrid mixture** as "a mixture of a flammable gas with either a combustible dust or a combustible mist." Hybrid mixtures may occur in, for instance, coal mining, spray drying of solids, or dumping solid particles into a vessel containing flammable solvents. Hybrid mixtures can also arise when particles are processed with a flammable solvent and later dried—during storage residual solvent is released and a hybrid mixture is formed.

Bartknecht (1981) discusses hybrid mixtures and available experimental data. He draws the following conclusions:

1. Dust–air and gas–air mixtures that are not flammable by themselves may combine to form a flammable mixture
2. The ignition energies for the hybrid mixtures may be reduced from that of the pure components and is particularly hazardous with readily ignited dusts.
3. The explosion violence of combustible dusts [as measured by P_{max} and $(dP/dt)_{max}$] can increase strongly with rising flammable gas concentration, possibly increasing the K_{St} value enough to move the mixture to a higher St class.

2.6. Kinetics and Thermochemistry

The combustion of a gas can be represented by a single overall stoichiometric reaction. For methane the overall reaction is

$$CH_4 + 2O_2 \rightarrow CO_2 + 2H_2O$$

For gaseous hydrocarbons that contain oxygen, nitrogen and sulfur a generic equation can be derived to describe the overall combustion reaction.

$$C_m H_x O_y N_u S_v + (m + x/4 - y/2 + v)\, O_2 \rightarrow mCO_2 + (x/2)H_2O + (u/2)N_2 + vSO_2$$

These reaction equations provide information on how the molecules and atoms are related stoichiometrically, but it does not tell us how they collide and combine. Furthermore, they assume that the combustion proceeds to stable compounds.

From a molecular basis, atoms and/or molecules collide and, if they have adequate energy, they will react. Elementary reactions are used to describe the actual collision and reaction mechanism. For methane, several dozen elementary reactions involving dozens of chemical species are used to represent the combustion process. An Arrhenius expression can be written for each reaction to represent the reaction rate. For instance, for the combustion of methane, the following elementary reaction

$$CH_4 + H \cdot \Leftrightarrow CH_3 \cdot + H_2$$

describes the collision of atomic hydrogen with a methane molecule to produce the methyl free radical and hydrogen.

The species $H \cdot$ and $CH_3 \cdot$ in the elementary reaction above are called free radicals, which only survive until they collide or react with another species. The concentrations of these unstable species are very low, typically 10^{-3} mole fraction or less. Most of the elementary reaction steps in a combustion process involve free radicals.

Energy is released during the combustion process. This energy heats the combustion products and any surrounding process equipment resulting in a higher temperature. If the combustion occurs in a closed vessel, the pressure will change due to the change in temperature and a change in the number of moles of gas determined by the reaction stoichiometry.

The complete numerical simulation of the combustion process is a very complex task. This requires an unsteady mass and energy balance coupled to the dozens of elementary reactions. In addition, the simulation requires a model for the turbulence induced by the rapidly generated and expanding combustion gases. These simulations provide the temperature, pressure and species concentrations as a function of distance and time. This is beyond the scope of this book.

An equilibrium calculation can be used to determine the final state of the combusting gases. This approach has become very popular in the past few years due to the availability of a number of equilibrium codes. These codes include CHEMKIN (available from Sandia Laboratories), STANJAN (available from W. C. Reynolds, Dept. of Mechanical Engineering, Stanford University) and EQS4WIN (available from Mathtrek Systems, Guelph, Ontario, Canada). An equilibrium code called Thermo Chemical Calculator (TCC) is also available for free on-line from Stanford University. The commercial consequence modeling package SuperChems (available from ioMosaic, Salem, NH) includes an equilibrium package.

These equilibrium codes typically provide the flexibility to set a number of parameters constant between the initial and final combustion state. If the combustion occurs in the open, such as an open flame, the pressure and enthalpy are identical between the initial and final states. The equilibrium code determines the final gas composition and temperature. If the combustion occurs in a closed vessel, then volume and internal energy are identical between the initial and final states. The equilibrium code determines the final temperature, pressure and gas composition.

In theory, the results of the equilibrium calculation should represent the final state of the combustion process after a very long period of time—a description of the transient heat, mass and kinetic processes is not required. This approach requires only thermodynamic data for all the chemical species, including products and perhaps the intermediate species. The detailed elementary reactions are not required. The procedure works as follows: given an initial mixture of gases at a specified temperature and pressure and a complete set of initial and product species, what is the final temperature, pressure and species concentrations that will minimize the Gibbs energy? The minimum Gibbs energy requirement is shown to correspond to the equilibrium case (Smith and Van Ness, 1987).

The equilibrium method requires the specification of a compete set of product species. This can present a problem since dozens of free radicals and other intermediate species are involved. Equilibrium code packages are capable of automatically selecting these species based on the elemental stoichiometric constraints. Thus, for methane, the user can specify the primary stable species of CH_4, O_2, H_2O and CO_2 and the code would select all the intermediate species possible based on atomic carbon, hydrogen and oxygen. The result would include species that would not be present in the actual combustion—the final numerical result would show a zero or very low concentration for these species. This procedure is effective as long as more species are specified than are actually present. Equilibrium codes today are capable of easily handling hundreds of species.

2.6.1. Calculated Adiabatic Flame Temperatures (CAFT)

The calculated adiabatic flame temperature (CAFT) is a powerful tool for estimating the flammability limits of complex gas mixtures (Hansel, Mitchell et al., 1991; Melhem, 1997). One procedure is to use a CAFT limit of 1200 K for both the upper and lower flammability limits—other values are used depending on whether more or less conservative results are desired. A study has shown (Mashuga and Crowl, 1999) that this approach works reasonably well for estimating the entire flammability zone on a flammability diagram (see Figure 2.7). These studies have also shown that the upper limit is estimated successfully only by including the intermediate unstable species using an equilibrium code. The lower limit can be estimated with or without the intermediate species.

Consider a gas mixture that is flowing through a well-insulated tube. At some point in the tube the gas mixture is ignited and a flame is stabilized. The process occurs at constant pressure. The energy of combustion is used to heat the combustion products. The final temperature is called the **adiabatic flame temperature**. This process occurs in the open, so the pressure is constant and the initial and final enthalpies are the same. Thus,

$$\Delta H = 0 \qquad (2\text{-}23)$$

where H is the total enthalpy (energy) and Δ denotes the final minus the initial state.

Since enthalpy is a state function, we can choose any path between the initial and final states. In this case it is convenient to choose a path that is comprised of two steps. First, the reaction occurs at the initial temperature to form products and release the energy of combustion. The second step involves heating of the products to the final adiabatic temperature. Combining these steps with Equation (2-23) results in

$$\Delta H = \Delta H_R{}^0 + \Delta H_P = n_R \Delta h_R + n_P \Delta h_P = 0 \qquad (2\text{-}24)$$

where $\Delta H_R{}^0$ is the total enthalpy change of reaction at the initial temperature (energy)

      ΔH_P is the total enthalpy change due to the heating of the products (energy)

      n_R is the total moles of reactants (moles)

      Δh_R is the molar enthalpy change of combustion (energy/mole)

      n_P is the total moles of product gases (moles)

      Δh_P is the molar enthalpy change due to heating of the products (energy/mole)

The molar enthalpy change due to the heating of the combustion products is determined from the heat capacity of the products,

$$\Delta h_P = \int_{T_1}^{T_{ad}} C_P^{\,prod} \, dT \qquad (2\text{-}25)$$

where $C_P^{\,prod}$ is the heat capacity of the product gases (energy/deg mole), T_1 is the initial temperature (deg), and T_{ad} is the adiabatic flame temperature (deg).

The heat capacity of the products, $C_P^{\,prod}$ is determined by

$$C_P^{\,prod} = \sum_{i=0}^{n} y_i C_{P_i} \qquad (2\text{-}26)$$

where y_i is the mole fraction of species i (moles) and C_{P_i} is the heat capacity for species i (energy/deg mole).

The heat capacities of the gases are, in general, a function of the temperature. Thus, the procedure to solve for the adiabatic flame temperature is trial and error, as follows:

1. Given: initial gas mixture composition, temperature and pressure; overall reaction stoichiometry; heat capacities of all species, including reactants and products; heat of combustion for the overall reaction. If the heat of combustion is not known, it can be determined from the heats of formation of the individual species.
2. Determine the final species composition. This usually assumes complete reaction until the limiting reactant is consumed. If oxygen is limiting, assign the oxygen to the carbon to make CO first, then H_2O and finally CO_2. Assume no nitrogen oxide formation.
3. Guess an adiabatic flame temperature.
4. Numerically integrate Equations (2-25) and (2-26) using a spreadsheet or other math package and the equations for the heat capacities for the individual species as a function of temperature.
5. Determine the temperature at which Equation (2-24) is satisfied.
6. Guess a new temperature based on the results of step 5 and return to step 4. Continue the procedure until a reasonably precise estimate of the temperature is obtained.

Most equilibrium codes are capable of determining the CAFT automatically. These codes would also determine the final gas composition, without specifying any reactions. Thus, step 2 in the above procedure would be completed by the code. The final gas composition dramatically affects the CAFT. The presence of a number of intermediate or unstable species, such as CO, NO·, H_2, H·, OH· and others, will affect the final temperature.

2.6.2. Example Application

The following example will illustrate an adiabatic combustion calculation assuming constant heat capacities. Also, it will be assumed that no intermediate or unstable combustion species are present in the final gas mixture and the reaction goes to completion.

EXAMPLE 2.6
Methane is combusted with a stoichiometric amount of air. The heat of combustion for methane at 298K is –212,800 cal/gm-mole. The heat capacities of the gases are assumed constant and are provided in the table below.

Species	Heat Capacity (cal/gm-mole °C)
CH_4	19.7
O_2	7.8
CO_2	9.8
H_2O	10.5
N_2	7.8

Determine the calculated adiabatic flame temperature (CAFT) for this system.

Solution: The procedure shown in Equations (2-23) through (2-26) is used to determine the CAFT. Since the heat capacities are constant with temperature, the solution is direct rather than by trial and error. Step 1 is already completed by the problem definition. The final species composition (Step 2) is determined assuming complete reaction until the limiting reactant is consumed. In this case, the fuel is provided in a stoichiometric mix and is consumed. The initial gas composition is:

Species	Moles	Mole Fraction, x_i
CH_4	1	0.0951
O_2	2	0.1901
CO_2	0	0.0
H_2O	0	0.0
N_2	7.52	0.7148
Total:	10.52	1.0000

The final gas composition is:

Species	Moles	Mole Fraction, x_i
CH_4	0	0.0
O_2	0	0.0
CO_2	1	0.0951
H_2O	2	0.1901
N_2	7.52	0.7148
Total:	10.52	1.0000

Step 3 in the procedure is not required since the solution is direct. For Step 4, Equation (2-25) is integrated directly since the heat capacities are constant. The heat capacity of the final mixture is given by Equation (2-26)

$$C_P^{prod} = y_{N_2} C_P^{N_2} + y_{CO_2} C_P^{CO_2} + y_{H_2O} C_P^{H_2O}$$
$$= (0.7148)(7.80 \text{ cal/gm-mole } °C) + (0.0951)(9.8 \text{ cal/gm-mole } °C)$$
$$+ (0.1901)(10.5 \text{ cal/gm-mole } °C)$$
$$= 8.50 \text{ cal/gm-mole } °C$$

Substituting into Equation (2-25),

$$\Delta h_P = \int_{298}^{T_{ad}} C_P^{prod} dT = C_P^{prod} \int_{298}^{T_{ad}} dT = (8.50 \text{ cal/gm-mol } °C)(T_{ad} - 298)$$

For Step 5, Equation (2-24) is used

$$n_R \Delta h_R = -n_P \Delta h_P$$

$$(1 \text{ mole})(-212,800 \text{ cal/gm-mole})$$
$$= -(10.52 \text{ moles})(8.503 \text{ cal/gm-mole } °C)(T_{ad} - 298)$$

Solving gives

$$T_{ad} = 2403°C = 2676 \text{ K}.$$

This example assumes constant heat capacities over the entire temperature range. If a more rigorous approach with variable heat capacities is used, the resulting temperature is 2328 K. If we further add CO to the species list and perform an equilibrium calculation, the answer is 2260 K. Finally, if we select C, O, H, and N as elements and select all 173 candidate species in the thermodynamic data base, and then perform an equilibrium calculation, the answer is 2226 K. In this case, CO, H_2, $OH\cdot$, $O\cdot$, $NO\cdot$, and $H\cdot$ are present in small amounts.

2.7. Gas Dynamics

An explosion results from the very rapid release of energy. The energy release must be sudden enough to cause a local accumulation of energy at the site of the explosion. This energy is then dissipated by a variety of mechanisms, including formation of a pressure wave, projectiles, thermal radiation, acoustic energy, or physical translation of equipment. The damage from an explosion is caused by the dissipating energy.

If the explosion occurs in a gas, the energy causes the gas to expand rapidly, forcing back the surrounding gas and initiating a pressure wave which moves rapidly outward from the blast source. The pressure wave contains energy which results in damage to the surroundings. For chemical plants, much of the damage from explosions is done by this pressure wave. Thus, in order to understand explosion impacts, the dynamics of the pressure wave must be understood. Many references are available in this area (Glasstone, 1962; Baker, 1983; Kinney and Graham, 1985; Lees, 1986; Baker, Cox et al., 1988; AIChE, 1994; Lees, 1996).

A pressure wave propagating in air is called a **blast wave**. If the pressure front has a very abrupt pressure change, as shown in Figure 2.19 as a function of distance at a fixed time, and in Figure 2.20 as a function of time at a fixed location, it is called a **shock wave** or **shock front**. A shock wave is expected from high explosive materials, such as TNT, but it can also occur from the sudden rupture of a pressure vessel. Figures 2.19 and 2.20 show the typical abrupt rise in pressure at the shock front, followed by a decrease in pressure behind it. The maximum pressure over ambient is called the **peak overpressure**.

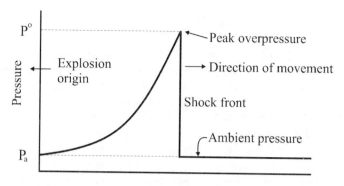

Distance from explosion origin

Figure 2.19. A typical shock wave at a fixed time.

The definitions for shock and blast waves in the literature and common practice vary between reference sources. It is clear that a shock wave involves a very abrupt, almost instantaneous, change in pressure that results from the explosion of a material such as TNT. A blast wave, on the other hand, is used more generally to include both shock waves and pressure waves that do not have an abrupt pressure change, such as would occur from the combustion of a flammable gas.

Figure 2.20 shows the variation in pressure with time for a typical shock wave at a fixed location some distance from an explosion site in the open. The explosion occurs at time t_0. There exists a small but finite time, t_1, before the shock front travels from its explosive origin to the affected location. The time, t_1, is called the **arrival time**. At t_1, the shock front has arrived and a peak overpressure is observed immediately followed by a strong transient wind. The pressure quickly decreases to ambient at t_2 but the wind continues in the same direction as the shock wave for a short time. The time period t_1 to t_2 is called the overpressure or positive phase dura-

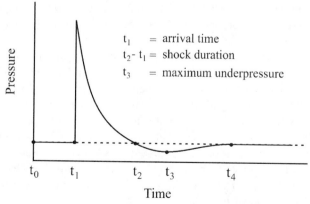

Figure 2.20. Shock wave pressure at a fixed location.

tion. The overpressure duration is typically (but not always) the period of greatest destruction to free standing structures so its value is important for estimating damage. The decreasing pressure continues to drop below ambient to a maximum underpressure at t_3. For most of the underpressure period (also called the negative pressure period) from $t_2 - t_4$ the blast wind reverses direction and flows toward the explosion origin as required to conserve mass. There is some damage associated with the underpressure period, but since the maximum underpressure is only a few psi for typical explosions, the damage is much less than that of the overpressure period. The underpressure for very large explosions and nuclear explosions can be quite large, however, resulting in considerable damage. Lower pressure vapor cloud deflagrations and bursting pressure vessels can have an underpressure of comparable magnitude to the overpressure (Tang and Baker, 1999). After attaining the maximum underpressure at t_3, the pressure will approach ambient at t_4. At this time the blast wind and the direct destruction have terminated.

An important consideration is how the pressure is measured as the blast wave passes. If the pressure transducer is at right angles to the blast wave, the overpressure measured is called the **side-on overpressure** (sometimes called the **free field overpressure**). This is the overpressure shown in Figures 2.19 for a shock wave. At a fixed location, shown in Figure 2.20, the side-on overpressure increases abruptly to its maximum value (**peak side-on overpressure**) and then drops off as the blast wave passes. If the pressure transducer is placed in the middle of a large wall facing toward the oncoming shock wave, then the pressure measured is the **reflected overpressure**. The reflected overpressure is a manifestation of the shock wave pressure and the dynamic pressure of the blast wind acting on an obstacle in their path. The **dynamic pressure** is defined as $\frac{1}{2}\rho u^2$, where ρ is the gas density and u is the gas velocity. The reflected pressure for low side-on overpressures is about twice the side-on overpressure and can reach as high as 8 or more times the side-on overpressure for strong shocks (Baker, Cox et al., 1988). The reflected overpressure is a maximum when the blast wave arrives normal to the wall or object of concern and decreases as the angle changes from normal. References often report overpressures without specifying whether they are side-on or reflected. Care should be exercised in using such information.

A time trace for both side-on and dynamic overpressure recorded at a fixed location for a shock wave is shown in Figure 2.21. The relative magnitudes of the side-on and dynamic pressures are a function of the side-on overpressure; at side-on overpressures of less than about 70 psi the dynamic pressure is less than the side-on overpressure. The maximum value of the dynamic pressure is called the **peak dynamic pressure**. The peak dynamic pressure decreases with increasing distance from the explosion similar to peak shock overpressure, but at a different rate. At a fixed location the dynamic pressure behaves similar to side-on overpressure as a function of time. However the pressure decay after the shock is often different, as shown in Figure 2.21. The dynamic pressure decays to zero

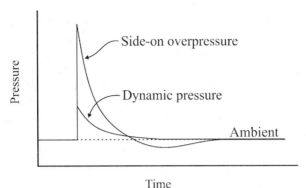

Time

Figure 2.21. *Side-on overpressure and dynamic pressure at a fixed location. For side-on overpressures of less than about 70 psi the dynamic pressure is less than the side-on overpressure, as shown.*

sometime later than the side-on overpressure due to the momentum of the moving air following behind the shock front

As the blast wave moves out from the explosion origin, energy is dissipated and the blast wave decays. However, the volume affected increases as the cube of the distance from the blast. Thus, it is important to understand the relationship between the overpressure and the distance from the explosion. Figure 2.22 shows a blast wave at various times as it moves outward from its explosive origin. As it proceeds the peak overpressure at the front decreases and the duration of the blast wave increases. The last curve (t_6) in Figure 2.22 shows an underpressure where a partial vacuum is produced. This phase is somewhat longer than its positive phase and has a reversal in wind direction.

The blast **impulse** is defined as the change in momentum and has dimensions of force-time product. For a blast wave, the area under the pressure–time curve is

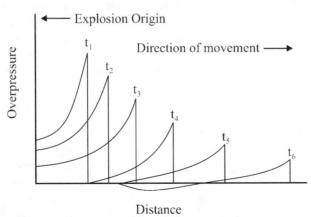

Distance

Figure 2.22. *Overpressure curves at various times after the initial explosion.*

the impulse per unit of projected area (Kinney and Graham, 1985). The impulse is reported separately for each of the overpressure and underpressure periods.

Damage from blast waves depend on many factors, including the shape, duration and magnitude of the blast wave; the standoff distance between the explosion and a receptor; the orientation and size of the receptor and the type of construction of the receptor. Damage estimates almost always require estimates of peak overpressure and impulse. It is generally recognized that simplistic damage estimates based on side-on overpressure only are adequate for structural damage predictions (Baker, Cox et al., 1988). More detail on this is provided in Section 2.13.2.

2.7.1. Detonations and Deflagrations

The difference between a detonation and deflagration depends on whether the reaction front propagates above or below the speed of sound in the unreacted material. For ideal gases, the speed of sound or sonic velocity is a function of temperature and molecular weight only and has a value of 344 m/s (1129 ft/s) for air at 20°C. Fundamentally, the sonic velocity is the speed at which information is transmitted through a gas.

In some reactions the reaction front is propagated by a strong pressure wave which compresses the unreacted material ahead of the reaction front, raising its temperature above its autoignition temperature. This compression occurs very rapidly, resulting in an abrupt pressure change or shock in front of the reaction front. This is classified as a detonation, resulting in a reaction front and leading shock wave which propagates into the unreacted mixture at or above the sonic velocity.

For a deflagration, the energy from the reaction is transferred to the unreacted mixture by heat conduction and molecular diffusion. These processes are relatively slow, causing the reaction front to propagate at a speed less than the sonic velocity.

Figure 2.23 shows the physical differences between a detonation and a deflagration for a combustion reaction which occurs in the gas phase in the open. For a detonation the reaction front moves at a speed greater than the speed of sound. A shock front is found a short distance in front of the reaction front. The reaction front provides the energy for the shock front and continues to drive it at sonic or greater speeds.

For a deflagration, the reaction front propagates at a speed less than the speed of sound. The pressure front moves at the speed of sound in the unreacted gas and moves away from the reaction front.

The pressure fronts produced by detonations and deflagrations in the open are markedly different, as shown on Figure 2.23. A detonation produces a shock front with an abrupt pressure rise. The maximum pressure depends on the phase and type of reacting material. Gaseous detonations are in the 15–20 atm range, whereas condensed phase materials may exceed 100 kbar. The duration depends on the explosion energy and can be milliseconds to tens of milliseconds. The pres-

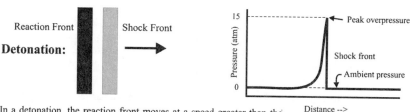

In a detonation, the reaction front moves at a speed greater than the speed of sound, driving the shock front immediately preceding it. Both fronts move at the same speed.

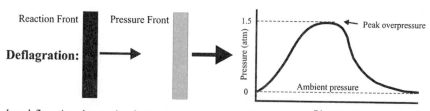

In a deflagration, the reaction front moves at a speed less than the speed of sound, while the pressure front moves away from the reaction front at the speed of sound.

Figure 2.23. *Comparison of detonation and deflagration gas dynamics. The explosion is initiated to the far left.*

sure front resulting from a deflagration in the open is characteristically long (milliseconds to hundreds of milliseconds in duration), broad and smooth (without an abrupt shock front) and with a maximum pressure typically one or two atmospheres. The pressure values shown on Figure 2.23 are only used to demonstrate the difference between a detonation and deflagration—actual values will vary considerably.

The behavior of the reaction and pressure fronts will differ from those shown in Figure 2.23 depending on the local geometry constraining the fronts. Different behavior will occur if the fronts propagate in a closed vessel, a pipeline, or through a congested process unit. The particular case for propagation within a closed sphere is presented in Figures 2.4 and 2.5 and discussed in Section 2.1. The gas dynamic behavior for more complex geometries is beyond the scope of this text.

Figure 2.24 shows a deflagration wave propagating through air. The higher pressure part of the curve has a higher temperature and hence a greater speed than the lower pressure parts. Thus, the higher pressure part of the curve overtakes the forward lower pressure part, resulting in a steeper front part of the curve as the pressure wave propagates away from the blast. This process may continue until the pressure wave approaches the shape of a detonation blast wave. This process is called "shocking up."

The damage due to either a detonation or deflagration is typically different. The damage produced by a detonation of the same energy is typically greater than

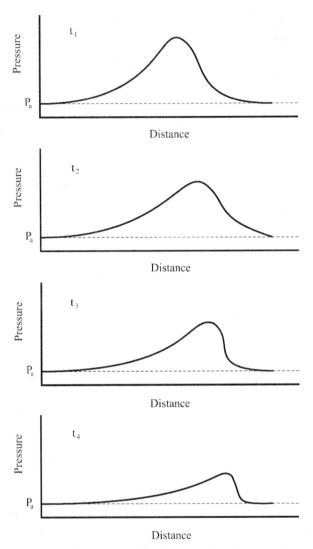

Figure 2.24. *A sequence of pressure profiles for increasing times, showing how a defla-gration wave approaches the shape of a detonation.*

the damage from a deflagration, primarily due to the much higher peak overpressure for the detonation, although this is a simplification of a complex situation. A deflagration should not be discounted due to the lower peak pressure—the lower peak pressure is offset by the much longer duration of the pressure wave which may be more damaging for certain structural components. This subject is discussed in more detain in Section 2.13.2.

Consider the specific example of combustion occurring within a process vessel resulting in vessel rupture and explosion. If the explosion is due to a deflagration, only a few vessel pieces would be expected. Also, the wall thickness will be thinner near the tear lines as the vessel wall is deformed just prior to rupture—this is called stress fracture or ductile failure. If the explosion is due to a detonation, many vessel pieces result since the vessel wall shatters due to the abrupt arrival of the shock front. The vessel pieces are also not thinned near the tear lines since inadequate time is available prior to rupture for wall thinning—this is called brittle fracture. The difference in the two failure modes is a consequence of the rate at which the explosion pressure is applied to the vessel wall. In the case of a deflagration, the pressure is applied slowly and the wall has time to stretch, resulting in a tearing failure. For a detonation, the pressure is applied so abruptly that the wall does not have time to stretch before it fails.

2.7.2. Estimating Peak Side-on Overpressures

For shock waves resulting from the detonation of high explosives, a well-defined relationship exists between overpressure and mass of TNT detonated. Figure 2.25 shows the scaled peak side-on overpressure and scaled impulse correlated as a function of the scaled distance Z, which is defined as,

$$Z = \frac{R}{W^{1/3}} \tag{2-27}$$

where Z is the scaled distance (distance/mass$^{1/3}$), R is the distance from the explosion center (distance), and W is the mass of TNT (mass).

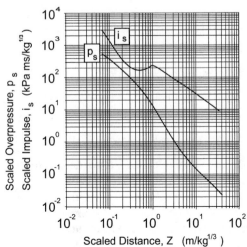

Figure 2.25. *Scaled overpressure and inpulse curves for a TNT explosion on a surface (Lees, 1996).*

Figure 2.25 is in metric units with the mass in kilograms and the distance in meters.

The scaled overpressure is defined as

$$p_s = \frac{p_0}{p_a} \tag{2-28}$$

where p_s is the scaled overpressure (unitless), p_0 is the peak side-on overpressure (pressure), and p_a is the absolute ambient pressure (pressure). The term "overpressure" always refers to a gauge pressure.

The scaled overpressure data in Figure 2.25 are for a TNT explosion on the ground at sea level. Overpressure correlations are frequently presented for free-air explosions, with a distant ground surface. To convert surface data to free air data, the mass of TNT is halved to account for the loss of energy reflected by the ground surface.

The impulse and peak side-on overpressure are used to estimate damage for a number of cases. More detail on this is provided in Section 2.13.2.

2.7.3. Example Applications

EXAMPLE 2.7
Estimate the peak side-on overpressure and impulse at a distance of 50 m from the explosion of 200 kg of TNT. The ambient pressure is 101.3 kPa.

Solution: The scaled distance is determine from Equation (2-27),

$$Z = \frac{R}{W^{1/3}} = \frac{50 \text{ m}}{(200 \text{ kg})^{1/3}} = 8.55 \text{ m}/\text{kg}^{1/3}$$

From Figure 2.25, this corresponds to a scaled overpressure of 0.18. Thus, from Equation (2-28), the peak side-on overpressure is,

$$p_0 = p_s p_a = (0.18)(101.3 \text{ kPa}) = 18.3 \text{ kPA}$$

The scaled impulse is read directly from Figure 2.25 and has a value of 32 kPa-ms/kg$^{1/3}$. The impulse is then

$$(32 \text{ kPa-ms/kg}^{1/3})(200 \text{ kg})^{1/3} = 187 \text{ kPa-ms}$$

This overpressure might not appear to be significant, but it is enough to cause injuries and extensive damage to some types of buildings.

EXAMPLE 2.8
Estimate the distance to a peak side-on overpressure of 21 kPa for the explosion of 200 kg of TNT. The ambient pressure is 101.3 kPa.

Solution: The problem is the reverse of Example 2.8. The scaled overpressure is determined using Equation (2-28)

$$p_s = \frac{p_o}{p_a} = \frac{21 \text{ kPa}}{101.3 \text{ kPa}} = 0.207$$

From Figure 2.25 this corresponds to a scaled distance, Z, of 8.0 $\text{m/kg}^{1/3}$. The distance is determined from Equation (2.7).

$$R = ZW^{1/3} = (8.0 \text{ m/kg}^{1/3})(200 \text{ kg TNT})^{1/3} = 46.8 \text{ m}$$

EXAMPLE 2.9
Estimate the mass of TNT exploded if an overpressure of 10 kPa is measured at a distance of 100 m from the blast origin. The ambient pressure is 101.3 kPa.
 Solution: The scaled overpressure is determined using Equation (2-28)

$$p_s = \frac{p_o}{p_a} = \frac{10 \text{ kPa}}{101.3 \text{ kPa}} = 0.0987$$

From Figure 2.25 this corresponds to a scaled distance, Z, of 12 $\text{m/kg}^{1/3}$. The mass of TNT is determined from Equation (2-27),

$$W = \left(\frac{R}{Z}\right)^3 = \left(\frac{100 \text{ m}}{12 \text{ m/kg}^{1/3}}\right)^3 = 579 \text{ kg TNT}$$

2.7.4. Pressure Piling and Deflagration to Detonation Transition

Hydrocarbon combustion in an enclosed space can theoretically result in a pressure increase of from 6 to 10 times the initial absolute pressure (averaging about 8 P_o) (NFPA 68, 2002). However, if a combustion occurs in interconnected vessels, as shown in Figure 2.26, **pressure piling** may result. This occurs if the combustion is a deflagration and is due to the pressure front moving much faster than

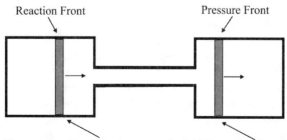

The reaction front here increases the initial pressure here prior to the arrival of the reaction front.

Figure 2.26. Pressure piling in interconnected vessels.

the reaction front. The pressure front will arrive in the connected vessel of Figure 2.26 prior to the reaction front. The pressure front will increase the pressure ahead of the reaction front (about 8 P_o) resulting in a final pressure after the reaction front passes of about $8 \times 8\ P_o = 64\ P_o$.

A deflagration can change into a detonation by a process called **deflagration to detonation transition** (**DDT**). It is more common in pipes and channels than in open spaces or vessels. Due to the geometric nature of pipes and channels, energy from the deflagration can accumulate in the pressure wave. If enough energy accumulates, the resulting adiabatic compression of the gas might lead to autognition and initiation of a detonation.

Chemical plants are complex processes with interconnected pipes, vessels, tanks, valves and so forth. If a combustion is initiated inside the process, the combustion may begin as a deflagration. However, the deflagration may propagate down pipelines and through other process equipment, with the possibility of a transition to a detonation. Flame front acceleration can also occur outside of process pipes and equipment if the volume has sufficient congestion and confinement which results in flame speed acceleration due to combustion generated turbulence. See Grossel (2002) for more details on pressure piling and DDT.

2.8. Physical Explosions

A pressure vessel contains stored energy due to its internal pressure and represents an explosion hazard. If the vessel is overpressured beyond its mechanical strength, or the vessel integrity is lost, the energy is released suddenly and significant damage can result. The damage is caused by the pressure wave from the sudden gas release which propagates rapidly outward from the vessel. This pressure wave may be a shock wave, depending on the nature of the failure. Flying fragments from the vessel wall or structure can also cause damage. If the vessel contents are flammable, a subsequent fire or vapor cloud explosion might result.

The energy contained in the compressed gas within the vessel is used to (1) stretch and tear the vessels walls, (2) provide kinetic energy to the fragments, and (3) provide energy for the pressure wave. Some of the energy of expansion also becomes "waste" thermal energy which is not concentrated enough to cause any thermal damage.

If the vessel contains all gas, the energy for the pressure wave is derived from the rapid expansion of the released gas. If the vessel contains a liquid with a pressurized vapor space, then the result is dependent on the liquid temperature. If the vessel contains both gas and liquid and the liquid is below its normal boiling point temperature, then the pressure energy is derived from the rapid expansion of the vapor space gases—the liquid remains unchanged and drains out. If the vessel contains both gas and liquid and the liquid is stored under pressure at a temperature above its normal boiling point, then the pressure wave may derive additional energy from the rapid flashing of a portion of the liquid. See Section 2.8.1.

There are four methods used to estimate the energy of explosion for a pressurized gas: Brode's equation, isentropic expansion, isothermal expansion and thermodynamic availability. Only Brode's equation will be presented here with the other three methods presented in Appendix B. Appendix B also contains examples on how to use these equations.

Brode's method (Brode, 1959) is perhaps the simplest approach. It determines the energy required to raise the pressure of the gas at constant volume from atmospheric pressure to the final gas pressure in the vessel. The resulting expression is

$$E = \frac{(P_2 - P_1)V}{\gamma - 1}$$
(2-29)

where E is the energy of explosion (energy)

P_2 is the burst pressure of the vessel (force/area)

P_1 is the ambient pressure (force/area)

V is the volume of the vessel (volume)

γ is the heat capacity ratio for the gas (unitless)

A portion of the potential explosion energy of vessel burst is converted into kinetic energy of the vessel pieces and other inefficiencies (such as strain energy in the form of heat in the vessel fragments). For estimation purposes, it is not uncommon to subtract 50% of the total potential energy to calculate the blast pressure effects from vessel burst.

Physical explosions are discussed in greater detail elsewhere, with curves provided to estimate overpressures at a fixed distance from a bursting pressure vessel (Baker, Cox et al., 1988; AIChE, 1994, 1999a).

2.8.1. BLEVEs

A **BLEVE**, or **boiling liquid expanding vapor explosion**, occurs when a vessel containing liquid above its normal boiling point fails catastrophically. The rapid depressurization during vessel failure results in sudden flashing of part of the liquid into vapor. The damage is caused, in part, by the pressure wave from the rapid flashing of the liquid and expansion of the vapor. Projectile damage from the container pieces and impingement damage from ejected liquids and solids is also possible. Vessel failure can result from causes such as an external fire, mechanical impact, corrosion, excessive internal pressure or metallurgical failure.

Figure 2.27 shows a vessel exposed to an external fire. The vessel walls below the liquid level are protected by heat transfer from the wall to the liquid, keeping the wall temperature low and maintaining the wall strength and structural integrity. However, the vessel walls above the liquid level are not protected due to poor heat transfer between the metal wall and the vapor, resulting in an increase in wall tem-

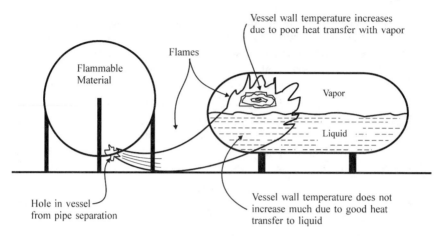

Figure 2.27. *The mechanism of BLEVE formation due to external fire. The metal wall exposed to vapor will eventually fail at a pressure below the vessel's rated pressure.*

perature and eventual structural failure due to loss of metal strength. The failure may occur at a vessel pressure well below the rated pressure of the vessel or the set pressure of the relief system. After the vessel fails, the liquid flashes almost instantaneously into vapor, resulting in a pressure wave and a vapor cloud. If the liquid is flammable and is ignited, a large fireball will form. The damage effects for this case include the overpressure from the vessel failure, thermal radiation and possible flame impingement from the fireball. BLEVEs can occur with vessels containing noncombustible liquids such as water. The damage effects for this case are due to the overpressure from the vessel failure, fragments from the vessel and impingement of the escaping hot water.

Films of actual BLEVE incidents involving flammable liquids clearly show several stages of BLEVE fireball development and growth. At the beginning of the incident, a fireball is formed quickly due to the rapid ejection of flammable material upon depressurization. This is followed by a much slower rise of the fireball due to buoyancy of the heated gases. A significant part of the damage from these types of BLEVEs is due to thermal radiation from the large, rising fireball, which can reach 100 m or more in diameter, rise several hundred meters in height, and last as long as 30 seconds. Blast or pressure effects from BLEVEs are generally limited to the near field (see Sections 2.9.1 to 2.9.3). Projectiles from the exploding vessel can be a significant hazard and can result in damage or involvement of adjacent units.

BLEVEs can happen quite quickly, with vessel failure occurring as soon as five minutes after initial fire exposure. In some instances BLEVEs have been quite delayed, with failure being many hours and sometimes days after initial fire exposure.

Additional information on BLEVEs is provided elsewhere (AIChE, 1994; Lees, 1996). Methods are available to estimate BLEVE fireball size, height and duration and to determine thermal radiation effects (AIChE, 1999a).

2.8.2. Rapid Phase Transition Explosions

A **rapid phase transition explosion** occurs when a liquid or solid undergoes a very rapid change in phase. If the phase change is from liquid to gas, or from solid to gas (sublimation), the volume of the material will increase hundreds or thousands of times, frequently resulting in an explosion. This is the process that causes popcorn to pop when the moisture within the kernel changes phase and expands rapidly.

The most common rapid phase transition explosion is due to the sudden exposure of a material to another material which is at a high enough temperature to cause the phase change. Key factors in rapid phase transition explosions include a large temperature difference, a boiling point for the material changing phase that is much lower than the temperature of the heat source, a large difference in heat capacity and a large area of contact between the material and the heat source.

Figure 2.28 illustrates a case history showing how this may occur. A hot oil at a temperature of 250°C was pumped into a distillation column for processing. Initially, valves A and B were closed. Due to a previous maintenance operation, water was present in the blocked-off pipe section between valves A and B. Valve A was accidently opened during the operation, exposing the water to the high temperature oil. The water flashed explosively, resulting in extensive internal damage to the column.

In any situation where a relatively low boiling point material may be brought into contact with a high temperature material there should be an evaluation of the potential for a rapid phase transition explosion.

Column

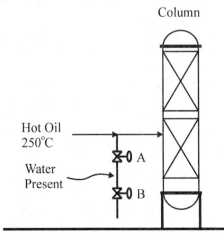

Figure 2.28. *The opening of Valve A exposed the water to the high-temperature oil, resulting in an explosion causing severe damage to the column.*

2.9. Vapor Cloud Explosions

A **vapor cloud explosion**, or **VCE**, occurs when a large quantity of flammable vapor or gas is released, mixes with air and is subsequently ignited. The vapor or gas fuel is usually released due to the loss of process containment. This could include the failure of a pipe, storage vessel or a process reactor. The rapid discharge of flammable process material through a relief system may also result in a VCE. The vapor may also originate from liquid stored under pressure to maintain it in the liquid state—the discharged liquid will flash rapidly into vapor at ambient pressure. The resulting explosion produces an overpressure which propagates outward from the explosion site as a blast wave. Significant damage from the resulting fireball is also possible due to thermal radiation.

Additional information on VCEs is provided elsewhere (AIChE, 1994; Lees, 1996; AIChE, 1999b).

Several conditions must generally be present for a VCE to result in damaging overpressure.

- The released material must be flammable.
- A cloud of sufficient size must form prior to ignition. If the cloud is too small, or is ignited early in the release, only a small fireball will result without significant overpressures. A jet or pool fire may subsequently form.
- The vapor cloud must mix with air to produce a sufficient mass in the flammable range of the material released. Without sufficient air mixing, a diffusion controlled fireball may result without significant overpressures developing.
- The speed of the flame propagation must accelerate as the vapor cloud burns. Without this acceleration, only a **flash fire** will result, which may produce significant damage due to thermal radiation and direct flame impingement.

These conditions are illustrated in Figure 2.29.

To complicate matters, the above quantities are related to each other. For example, the method and quantity of release will affect the degree of mixing with air and the size of the cloud. These interrelated qualitative features make the characterization of vapor cloud explosions very difficult.

Flame acceleration is an important part of the vapor cloud explosion. The flame accelerates when turbulence stretches and tears the flame front increasing its surface area. The primary turbulence sources are flow turbulence established in the unburned gas as it flows ahead of the flame front, pushed by the expanding combustion products behind it; and turbulence caused by the interactions of the gas with obstacles it encounters. In either case the turbulence results from the motion of the gas. As the turbulence increases, stretching the flame front, the rate at which the fuel is combusted increases because the area of the stretched and torn flame front has increased. As the rate of combustion increases the push on the unburned gases increases, causing them to move even faster, increasing the turbulence fur-

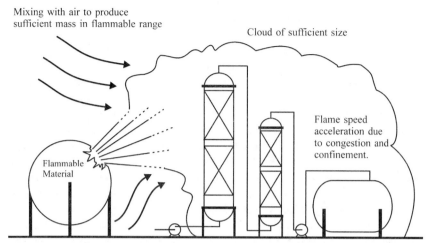

Mixing with air to produce
sufficient mass in flammable range

Cloud of sufficient size

Flame speed
acceleration due
to congestion and
confinement.

Flammable
Material

Figure 2.29. Conditions required for a vapor cloud explosion. An ignition source is also required, but these are abundant.

ther. This creates a feedback mechanism that accelerates the flame speed. Turbulence from the release could be a contributing factor near the point of release, but flame front acceleration to damaging overpressures can occur even in initially quiescent gas in a congested or confined process area. A congested process area, populated with pipes, pumps, valves, vessels, and other process equipment is adequate to result in significant flame speed acceleration.

In the past, a VCE was commonly called an unconfined vapor cloud explosion (UVCE). However, as discussed above, the degree of confinement significantly affects the behavior of a VCE. The combustion of a vapor completely in the open typically will result in a flash fire without significant overpressures. Thus, the word "unconfined" has been dropped from the terminology.

Most VCEs result in deflagrations—detonations are unlikely. Increasing levels of confinement and congestion increase the explosion overpressure. While vapor cloud detonations are rare, the higher overpressures can approach the severity of detonations.

Lenoir and Davenport (1992) have summarized over 200 VCE incidents. All (with possibly one exception) were deflagrations rather than detonations. They found that VCEs accounted for 37% of the number of property losses in excess of $50 million (corrected to 1991 dollars) and accounted for 50% of the overall dollars paid in insurance claims. Of the ten largest property losses in the process industry, seven were due to VCEs.

The major concerns for anyone involved with risk assessment related to vapor cloud explosions is the overpressure and/or impulse as a function of distance from the explosion. Once these are known the damage effects can be estimated. Additional detail on damage estimates is provided in Section 2.13.

Some methods to determine the overpressure/impulse require an estimate of the flammable mass. Estimation of the total flammable mass within the vapor cloud is difficult. A detailed monograph on this complex subject is provided elsewhere (AIChE, 1999b). Ideally, a dispersion model could be used to determine the flammable mass in the vapor cloud that is between the upper and lower flammability limits. Dispersion modeling of flammable gases is complex and the models have not been validated for congested volumes. At flammable concentrations the cloud is typically dense. The dense gas problem is three dimensional, requiring numerical or analytical integration of the concentration profiles within the cloud. It is also not clear what concentration to use for the extent of cloud combustion. The dispersion models do not account for variations in instantaneous cloud concentrations which may result in nonuniform burning of the vapor cloud. Furthermore, there is some evidence that a fuel–air mixture will burn beyond the flammability limits by overdriving it with an energetic ignition source. As a result of these difficulties, some risk analysts use a conservative concentration limit to define the combustible cloud equal to one-half of the lower flammability limit (AIChE, 1999a, 2000). Dispersion modeling is discussed further in several CCPS books (AIChE, 1996a; Hanna and Britter, 2002).

Four methods are available to estimate the overpressure and/or impulse as a function of distance from the explosion. These methods are: TNT equivalency, TNO multi-energy, Baker–Strehlow, and computational fluid dynamics (CFD). Details and examples on the application of these methods are provided elsewhere (AIChE, 1994, 1999a, 2000).

2.9.1. TNT Equivalency

TNT equivalency is a simple method for equating a known energy of a combustible fuel to an equivalent mass of TNT. This model has been described in many references (Robinson, 1944; Decker, 1974; Stull, 1977; Lees, 1986; Baker, Cox et al., 1988; AIChE, 1994; Lees, 1996; Crowl and Louvar, 2002). The approach is based on the assumption that an exploding fuel mass will behave like exploding TNT on an equivalent energy basis. The equivalent mass of TNT is estimated using the following equation,

$$W = \frac{\eta m E_c}{E_{TNT}}$$

(2-30)

where W is the equivalent mass of TNT (mass)
η is the empirical explosion efficiency (unitless)
m is the mass of flammable gas (mass)
E_c is the heat of combustion of the flammable gas (energy/mass)
E_{TNT} is the energy of explosion of TNT (energy/mass)

A typical value for the energy of explosion of TNT is 1100 cal/gm (4602 kJ/kg = 1980 Btu/lb). The energy of explosion for TNT is somewhat difficult to measure with a range of values reported.

The explosion efficiency represents one of the major problems in the equivalency method. The explosion efficiency is used to adjust the estimate for a number of factors, including incomplete mixing of the combustible material with air, incomplete conversion of the thermal energy to mechanical energy, and so forth. The explosion efficiency is empirical with most flammable cloud estimates varying between 1 and 10% as reported by a number of sources (Brasie and Simpson, 1968; Gugan, 1979; Lees, 1986; Lenoir and Davenport, 1992; Lees, 1996). Others have reported 5, 10, and 15% for flammable clouds of propane, diethyl ether, and acetylene (AIChE, 1994).

Although VCEs are most commonly deflagrations, the TNT equivalency method assumes a detonation and does not consider the effects due to flame acceleration from various levels of confinement and congestion. The TNT equivalency method also uses an overpressure curve that applies to point source detonations of TNT. As a result, the overpressure curve for TNT tends to overpredict the overpressure near the VCE, and to underpredict at distances away from the VCE. Finally, for large releases, a problem occurs in specifying the center of the explosion. Should the center be at the release point, the ignition source (if known), or the geometric center of the cloud? See Woodward (AIChE, 1999b) for much more discussion on these issues.

The advantage of the TNT equivalency method is that it is easy to apply due to the simple calculations.

The procedure to estimate the damage associated with an explosion using the TNT equivalency method is as follows:

1. Determine the total quantity of flammable material involved in the explosion. This can be estimated from the total quantity released or from a dispersion model (AIChE, 1999b).
2. Estimate the explosion efficiency and calculate the equivalent mass of TNT by Equation (2-30).
3. Use the scaling law given by Equation (2-27) and Figure 2.25 to estimate the peak side-on overpressure.
4. Use Tables 2.9 and 2.10 to estimate the damage for common structures and process equipment. Additional detail on damage estimates is provided in Section 2.13.2.

The procedure can be applied in reverse to estimate the quantity of material involved based on damage estimates.

2.9.2. TNO Multi-Energy Method

The TNO method identifies the confined volumes in a process, assigns a relative degree of confinement, and then determines the contribution to the overpressure

TABLE 2.9
Damage Estimates for Common Structures Based on Overpressure (Clancey, 1972)[a]

Pressure		Damage
psig	kPa	
0.02	0.14	Annoying noise (137 dB if of low frequency 10–15 Hz)
0.03	0.21	Occasional breaking of large glass windows already under strain
0.04	0.28	Loud noise (143 dB), sonic boom, glass failure
0.1	0.69	Breakage of small windows under strain
0.15	1.03	Typical pressure for glass breakage
0.3	2.07	"Safe distance" (probability 0.95 of no serious damage below this value); projectile limit; some damage to house ceilings; 10% window glass broken
0.4	2.76	Limited minor structural damage
0.5–1.0	3.4–6.9	Large and small windows usually shattered; occasional damage to window frames
0.7	4.8	Minor damage to house structures
1.0	6.9	Partial demolition of houses, made uninhabitable
1–2	6.9–13.8	Corrugated asbestos shattered; corrugated steel or aluminum panels, fastenings fail, followed by buckling; wood panels (standard housing) fastenings fail, panels blown in
1.3	9.0	Steel frame of clad building slightly distorted
2	13.8	Partial collapse of walls and roofs of houses
2–3	13.8–20.7	Concrete or cinder block walls, not reinforced, shattered
2.3	15.8	Lower limit of serious structural damage
2.5	17.2	50% destruction of brickwork of houses
3	20.7	Heavy machines (3000 lb) in industrial building suffered little damage; steel frame building distorted and pulled away from foundations.
3–4	20.7–27.6	Frameless, self-framing steel panel building demolished; rupture of oil storage tanks
4	27.6	Cladding of light industrial buildings ruptured
5	34.5	Wooden utility poles snapped; tall hydraulic press (40,000 lb) in building slightly damaged
5–7	34.5–48.2	Nearly complete destruction of houses
7	48.2	Loaded, lighter weight (British) train wagons overturned
7–8	48.2–55.1	Brick panels, 8–12 in. thick, not reinforced, fail by shearing or flexure
9	62.0	Loaded train boxcars completely demolished
10	68.9	Probable total destruction of buildings; heavy machine tools (7,000 lb) moved and badly damaged, very heavy machine tools (12,000 lb) survive
300	2068	Limit of crater lip

[a]These values should only be used for approximate estimates.

TABLE 2.10
Damage Estimates Based on Overpressure for Process Equipment (Stephens, 1970)[a]

Equipment	Overpressure, psi																								
	0.5	1.0	1.5	2.0	2.5	3.0	3.5	4.0	4.5	5.0	5.5	6.0	6.5	7.0	7.5	8.0	8.5	9.0	9.5	10	12	14	16	18	20
Control house steel roof	A	C	D				N																		
Control house concrete roof	A	E	P	D			N																		
Cooling tower	B			F			O																		
Tank: cone roof		D				K							U												
Instrument cubicle			A		I	LM						T													
Fired heater								I		T															
Reactor: chemical				G	A			I					P					T							
Filter				H					F										V		T				
Regenerator						I				IP					T										
Tank: floating roof						K							U								T				D
Reactor: cracking							I							I							T				
Pipe supports							P					SO													
Utilities: gas meter									Q																
Utilities: electronic									H	H					I			I		T					V
Electric motor																		I		T					
Blower										Q	R			T											
Fractionation column												PI						T							
Pressure vessel horizontal												I								V	T				
Utilities: gas regulator																				MQ					
Extraction column													I		I					V	T				
Steam turbine															I			I			M	S			V
Heat exchanger																I							T		
Tank sphere																					I	I	T		
Pressure vessel: vertical																					I	T			
Pump																					I		V		

[a]See the legend on the next page.

73

Legend to Table 2.10

A. Windows and gauges broken	L. Power lines are severed
B. Louvers fall at 0.2–0.5 psi	M. Controls are damaged
C. Switchgear is damaged from roof collapse	N. Block walls fall
D. Roof collapses	O. Frame collapses
E. Instruments are damaged	P. Frame deforms
F. Inner parts are damaged	Q. Case is damaged
G. Brick cracks	R. Frame cracks
H. Debris—missile damage occurs	S. Piping breaks
I. Unit moves and pipes break	T. Unit overturns or is destroyed
J. Bracing falls	U. Unit uplifts (0.9 tilted)
K. Unit uplifts (half tilted)	V. Unit moves on foundation

from this confined volume. Semi-empirical curves are used to determine the overpressure. Full details are provided elsewhere (TNO, 1997; Eggen, 1998; Mercx, A. C. van den Berg et al., 1998).

The basis for this model is that the energy of explosion is highly dependent on the level of confinement and congestion and less dependent on the fuel in the cloud.

The procedure for employing the multi-energy model to a vapor cloud explosion is given by the following steps (AIChE, 1994):

1. Perform a dispersion model to determine the extent of the cloud. Generally, this is performed assuming that equipment and buildings are not present, due to the limitations of dispersion modeling in congested areas. Several CCPS books discuss this issue (AIChE, 1999b; Hanna and Britter, 2002).

2. Conduct a field inspection to identify the congested areas. Normally, heavy vapors will tend to move downhill.

3. Identify potential sources of strong blast present within the area covered by the flammable cloud. Potential sources of strong blast include:
   - Congested areas and buildings such as process equipment in chemical plants or refineries, stacks of crates or pallets, and pipe racks;
   - Spaces between extended parallel planes, for example, those beneath closely parked cars in parking lots and open buildings in multistory parking garages;
   - Spaces within tubelike structures, for example, tunnels, bridges, corridors, sewage systems, culverts;
   - An intensely turbulent fuel–air mixture in a jet resulting from release at high pressure.

   The remaining fuel–air mixture in the cloud is assumed to produce a blast of minor strength.

4. Estimate the energy of equivalent fuel-air charges.
   • Consider each blast source separately.
   • Assume that the full quantities of fuel-air mixture present within the partially confined/obstructed areas and jets, identified as blast sources in the cloud, contribute to the blasts.
   • Estimate the volumes of fuel–air mixture present in the individual areas identified as blast sources. This estimate can be based on the overall dimensions of the areas and jets. Note that the flammable mixture may not fill an entire blast-source volume and that the volume of equipment may be considered where it represents an appreciable proportion of the whole volume.
   • Calculate the combustion energy E (in Joules) for each blast by multiplication of the individual volumes of the mixture by 3.5×10^6 J/m^3. This value is typical for the heat of combustion of an average stoichiometric hydrocarbon–air mixture.
5. Estimate strengths of individual blasts based on a scale from 1 to 10, with 10 being the strongest. Some companies have defined procedures for this; however, many risk analysts use their own judgment.
   • A safe and most conservative estimate of the strength of the sources of a strong blast can be made if a maximum strength of 10 is assumed—representative of a detonation. However, a source strength of 7 seems to more accurately represent actual experience. Furthermore, for side-on overpressures below about 0.5 bar, no differences appear for source strengths ranging from 7 to 10.
   • The blast resulting from the remaining unconfined and unobstructed parts of a cloud can be modeled by assuming a low initial strength. For extended and quiescent parts, assume minimum strength of 1. For more nonquiescent parts, which are in low-intensity turbulent motion, for instance, because of the momentum of a fuel release, assume a strength of 3.
6. Once the energy quantities E and the initial blast strengths of the individual equivalent fuel–air charges are estimated, the Sachs-scaled side-on overpressure and positive-phase duration at some distance R from a blast source is read from the blast charts in Figure 2.30 after calculation of the Sachs-scaled distance:

$$\bar{R} = \frac{R}{(E/P_a)^{1/3}} \tag{2-31}$$

where $\bar{R}$ is the Sachs-scaled distance from the charge (dimensionless)
   R is the distance from the charge (m)
   E is the charge combustion energy (J)
   P_a is the ambient pressure (Pa)

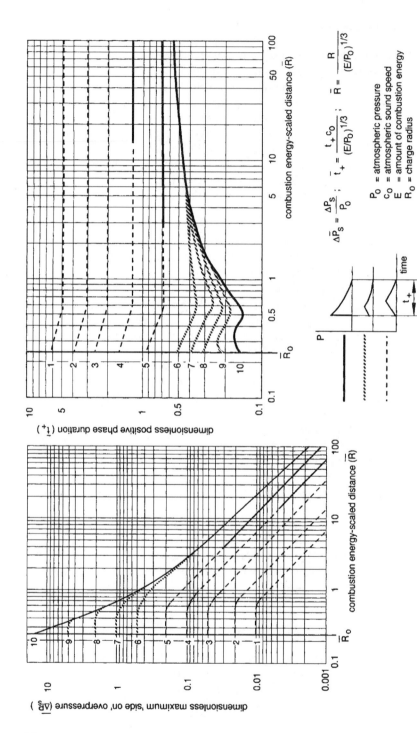

Figure 2.30. TNO multi-energy model for vapor cloud explosions (AIChE, 1994)

76

The blast peak side-on overpressure and positive-phase duration are calculated from the Sachs-scaled quantities:

$$P_s = \Delta \overline{P}_s \cdot P_a \qquad (2\text{-}32)$$

and

$$t_d = \overline{t}_d \left[\frac{(E/P_a)^{1/3}}{c_0} \right] \qquad (2\text{-}33)$$

where P_s is the side-on blast overpressure (Pa)
   $\Delta \overline{P}_s$ is the Sachs-scaled side-on blast overpressure (dimensionless)
   P_a is the ambient pressure (Pa)
   t_d is the positive-phase duration (s)
   $\overline{t}_d$ is the Sachs-scaled positive-phase duration (dimensionless)
   E is the charge combustion energy (J)
   c_0 is the ambient speed of sound (m/s)

If separate blast sources are located close to one another, they may be initiated almost simultaneously. In situations where it is not possible to rule out the events occurring simultaneously, the blasts should be superimposed. The most conservative approach to this issue is to assume a maximum initial blast strength of 10 and to sum the combustion energy from each source in question. Further definition of this important issue, for instance the determination of a minimum distance between potential blast sources so that their individual blasts may be considered separately, is a subject in present research.

If environmental and atmospheric conditions are such that vapor cloud dispersion can be expected to be very slow (during stable atmospheric conditions), the possibility of a vapor cloud detonation should also be considered if, in addition, a long ignition delay is likely. In that case, the full quantity of fuel mixed within detonable limits should be assumed for a fuel–air charge whose initial strength is maximum 10.

Three references (van Winderden, van den Berg et al., 1989; Kinsella, 1993; TNO, 1997) are available to assist the user in selection of the blast strength, the ignition energy and partial confinement. However, it is not clear how the results from each blast strength should be combined. The TNO method also assumes a ground level explosion.

2.9.3. Baker–Strehlow–Tang Method (AIChE, 1999a)

The Baker–Stehlow–Tang method is based on a flame speed, which is selected based on three factors: the reactivity of the released material, the flame expansion characteristics of the process unit (which relates to confinement and spatial configuration), and the obstacle density within the process unit. A set of semi-empirical curves is used to determine the overpressure.

This method is a modification of the original work by Strehlow (Strehlow, Luckritz et al., 1979), with added elements of the TNO multi-energy method. A complete description of the procedure is provided by Baker et al. (Baker, Tang et al., 1994; Tang and Baker, 1999).

In Strehlow's model a curve is selected based on flame speed, which affords the opportunity to use empirical data in the selection. The procedures from the TNO multi-energy method were adopted for determination of the energy term. Specifically, confinement is the basis of the determination of the size of the flammable vapor cloud that contributes to the generation of the blast overpressure. As in the TNO model, multiple blast sources can emanate from a single release.

Baker et al. (1994) state that experimental data suggest that the combined effects of fuel reactivity, obstacle density and confinement can be correlated to flame speed. They describe a set of 27 possible combinations of these parameters based on 2, 2.5 or 3D flame expansions (Baker, Doolittle et al., 1997). The results are shown in Table 2.11.

For 3D symmetry the flame is free to expand spherically from a point ignition source. The overall flame surface increases with the square of the distance from the point ignition source. The flame-induced flow field can decay freely in three directions. Therefore, flow velocities are low, and the flow field disturbances by obstacles are small.

In 2D symmetry, that is, a cylindrical flame between two plates, the overall flame surface area is proportional to the distance from the ignition point. Consequently, deformation of the flame surface will have a stronger effect than in the point-symmetry case.

The 2.5D symmetry was adopted for situations in which there is some form of confinement that is more restrictive than 3D but does not merit a 2D rating. A common situation where this applies is a light weight or low strength roof that creates 2D confinement, but the roof blows off during the explosion creating a vent. It is very common to find compressor shelters with a light roof that provides weather protection for maintenance. Another situation where 2.5D applies is a plane of very high congestion or a plane with partial 2D confinement. An example of this situation is a pipe rack with such a dense layer of pipes that 3D confinement is not conservative.

1D flame speeds are provided in Table 2.11 but this configuration is rarely encountered in actual plants. In 1D symmetry, for example, a planar flame in a tube, the projected flame surface area is constant. There is hardly any flow field decay, and flame deformation has a very strong effect on flame acceleration.

The flame speeds in Table 2.11 are expressed in Mach number units, with the Mach number defined as the local flow velocity divided by the local speed of sound. Note that the values in Table 2.11 represent the maximum flame speed for

TABLE 2.11
Flame Speed in Mach Number for Soft Ignition Sources (Baker, 2003)

1D Flame Expansion Case (not used)		Obstacle Density		
		High	Medium	Low
Reactivity	High	5.2	5.2	5.2
	Medium	2.27	1.77	1.03
	Low	2.27	1.03	0.294

2D Flame Expansion Case		Obstacle Density		
		High	Medium	Low
Reactivity	High	DDT[a]	DDT	0.59
	Medium	1.6	0.66	0.47
	Low	0.66	0.47	0.079

2.5D Flame Expansion Case		Obstacle Density		
		High	Medium	Low
Reactivity	High	DDT	DDT	0.47
	Medium	1.0	0.55	0.29
	Low	0.50	0.35	0.053

3D Flame Expansion Case		Obstacle Density		
		High	Medium	Low
Reactivity	High	DDT	DDT	0.36
	Medium	0.50	0.44	0.11
	Low	0.34	0.23	0.026

[a]DDT = Deflagration to detonation transition

each case and will produce a conservative result. The 3D flame speeds in Table 2.11 are the result of extensive testing conducted for a joint industry research program (Baker, 2003). The actual experimental 3D flame speed values have been scaled-up to account for the size of a typical plant. The 2D flame speeds in Table 2.11 are the higher of (1) the original 2D values provided in previous publications (Baker, Doolittle et al., 1997) and (2) recent experimental 2D flame speeds scaled-up to account for the size of a typical plant. For most of the cases, the scaled-up experimental value is lower than that provided in Table 2.11 giving a conservative result.

TABLE 2.12
Geometric Considerations for the Baker–Strehlow Vapor Cloud Explosion Model (Baker et al., 1994)

Dimension	Description	Geometry
3D	"Unconfined volume," almost completely free expansion	
2.5D	Compressor shelters with lightweight roofs; dense pipe racks	Between 3-D and 2-D
2D	Platforms carrying process equipment; space beneath cars; open-sided multistory buildings	
1D	Tunnels, corridors, or sewage systems	

Reactivity is classified as low, medium, and high according to the following recommendations of TNO (Zeeuwen and Wiekema, 1978). Methane and carbon monoxide are the only materials regarded as low reactivity, whereas only hydrogen, acetylene, ethylene, ethylene oxide, and propylene oxide were considered to be highly reactive. All other fuels are classified as medium reactivity. Fuel mixtures are classified according to the concentration of the most reactive component.

Obstacle density is classified as low, medium and high, as shown in Table 2.13, as a function of the blockage ratio and pitch. The blockage ratio is defined as the ratio of the area blocked by obstacles to the total cross-section area. The pitch is

TABLE 2.13
Confinement Considerations for the Baker–Strehlow Vapor Cloud Expansion Model (Baker et al., 1994)

Type	Obstacle Blockage Ratio per Plane	Pitch for Obstacle Layers	Geometry
Low	Less than 10%	One or two layers of obstacles	
Medium	Between 10% and 40%	Two to three layers of obstacles	
High	Greater than 40%	Three or more fairly closely spaced obstacle layers	

defined as the distance between successive obstacles or obstacle rows. There is a value for the pitch which results in maximum flame front acceleration: when the pitch is too large, the wrinkles in the flame front will burn out and the flame front will slow down before the next obstacle is reached. When the pitch is too small, the gas pockets between successive obstacles are relatively unaffected by the flow (Baker, Tang et al., 1994). Additional information is available elsewhere (Hanna and Britter, 2002).

Low density assumes few obstacles in the flame's path, or the obstacles are widely spaced (blockage ratio less than 10%), and there are only one or two layers of obstacles. At the other extreme, high obstacle density occurs when there are three or more fairly closely spaced layers of obstacles with a blockage ratio of 40% or greater per layer. Medium density falls between the two categories.

A high obstacle density may occur in a process unit in which there are many closely spaced structural members, pipes, valves, and other turbulence generators. Also, pipe racks in which there are multiple layers of closely spaced pipes must be considered high density.

Once the flame speed is determined from Table 2-11, then Figure 2.31 is used to determine the side-on overpressure and Figure 2.32 is used to determine the impulse of the explosion. The curves on these figures are labeled with the flame velocity, M_f. M_f denotes the flame velocity with respect to a fixed coordinate system, and is called the "apparent flame speed"—this is the value that should be used in predicting the scaled overpressure from Figure 2.31. The flame speed is

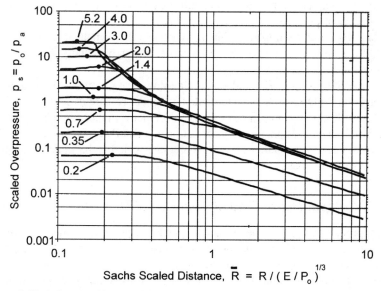

Figure 2.31. Baker–Strehlow model for vapor cloud explosions. This curve provides the scaled overpressure as a function of the Sachs scaled distance (Tang and Baker, 1999).

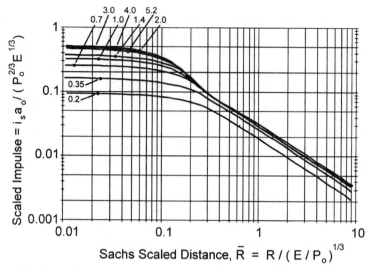

Figure 2-32. *Baker–Strehlow model for vapor cloud explosions. This curve provides the scaled impulse as a function of the Sachs scaled distance (Tang and Baker, 1999).*

expressed in Mach number, and is calculated in relation to the ambient speed of sound.

Figures 2.31 and 2.32 are based on free air bursts—for a ground or near ground level explosion, the energy is multiplied by a factor of two to account for the reflected blast wave.

The procedure for implementing the Baker–Strehlow method is similar to the TNO Multi-Energy method, with the exception that steps 4 and 5 use Table 2.11 and Figures 2.31 and 2.32.

The TNO multi-energy and Baker–Strehlow–Tang methods are analogous, although the TNO method tends to predict a higher pressure in the near field and the Baker–Strehlow–Tang method tends to predict a higher pressure in the far field. Both methods require more information and detailed calculations than TNT equivalency and both methods predict a pressure which is the same at any given distance from the blast.

2.9.4. Computational Fluid Mechanics (CFD) Method

This approach solves the coupled time-dependent equations of energy, mass and momentum transfer as the reaction and pressure fronts propagate. This approach requires a complete specification of the geometry and some definition of the combustible fuel concentration. The combustible fuel concentration is approached by assuming a simple stoichiometric cloud—more complicated clouds can be defined based on dispersion calculations. Sometimes it is necessary to use CFD to characterize the time-dependent cloud as it spreads out from the release point until it reaches an ignition point. Thus, the CFD approach requires much more detail, time and effort than the other methods, but can provide a more accurate prediction when properly used. The CFD method is becoming more popular due to the availability of CFD software and increased computer speeds.

2.9.5. Example Applications

EXAMPLE 2.10
Estimate the distance for a 3 psi side-on overpressure from a vapor cloud explosion (VCE) from the rupture of a holding tank containing 18 tons of hexane. Use the TNT equivalency method.

Solution: The scaled pressure is

$$p_s = \frac{3 \text{ psi}}{14.7 \text{ psi}} = 0.204$$

From Figure 2.25 this represents a scaled distance of $8.0 \text{ m/kg}^{1/3}$. Converting the units,

$$\frac{8.0 \text{ m/kg}^{1/3}}{(0.3048 \text{ m/ft})(1 \text{ lb}/0.4536 \text{ kg})^{1/3}} = 20.2 \text{ ft/lb}^{1/3}$$

The heat of combustion for hexane is found in Appendix A with a value of 4194.5 kJ/mol. The units are converted to:

$$\left(\frac{4194.5 \text{ kJ}}{\text{mol}}\right)\left(\frac{1 \text{ mol}}{86 \text{ gm}}\right)\left(\frac{453.6 \text{ gm}}{1 \text{ lb}}\right)\left(\frac{0.9484 \text{ Btu}}{1 \text{ kg}}\right) = 21,000 \text{ Btu/lb}$$

The energy of explosion for TNT is 1980 Btu/lb. Equation (2-30) is used with an explosion efficiency of 10%, although this value is uncertain and fairly arbitrary. The equivalent mass of TNT is then,

$$W = \frac{\eta m E_c}{E_{\text{TNT}}} = \frac{(0.10)(18 \text{ tons})(2000 \text{ lb/ton})(21,000 \text{ Btu/lb})}{1980 \text{ Btu/lb}} = 38,200 \text{ lb TNT}$$

Equation (2-27) is used to calculate the distance a 3 psi overpressure will be generated for the 38,200 lb equivalent mass of TNT.

$$R = ZW^{1/3} = (20.2 \text{ ft/lb}^{1/3})(38,200 \text{ lb})^{1/3} = 678 \text{ ft}$$

EXAMPLE 2.11
A VCE due to the rupture of a propane tank results in minor damage to houses reported 300 m away. Use the TNT equivalency method to estimate the amount of propane responsible for the damage.

Solution: From Table 2.9 the overpressure consistent with the explosion description is 4.8 kPa. This is a scaled overpressure of

$$p_s = \frac{4.8 \text{ kPa}}{101.3 \text{ kPa}} = 0.040$$

Using Figure 2.25 this overpressure corresponds to a scaled distance of 25 m/kg$^{1/3}$. From Equation (2-27) the total equivalent mass of TNT responsible for the reported damage is estimated.

$$W = \left(\frac{R}{Z}\right)^3 = \left(\frac{300 \text{ m}}{25 \text{ m}/\text{kg}^{1/3}}\right)^3 = 1728 \text{ kg}$$

Use an energy of explosion for TNT of 4602 kJ/kg and an explosion efficiency of 5% listed for propane. From Appendix A the heat of combustion for propane is 2220 kJ/mole or 50,452 kJ/kg. The mass of propane is estimated from Equation (2-30),

$$m = \frac{WE_{TNT}}{\eta E_c} = \frac{(1728 \text{ kg})(4602 \text{ kJ}/\text{kg})}{(0.05)(50,452 \text{ kJ}/\text{kg})} = 3150 \text{ kg propane}$$

EXAMPLE 2.12 (AIChE, 1999a)
Consider the explosion of a propane–air vapor cloud confined beneath a storage tank. The tank is supported 1 meter off the ground by concrete piles. The concentration of vapor in the cloud is assumed to be at stoichiometric concentrations. Assume a cloud volume of 2094 m^3, confined below the tank, representing the volume underneath the tank. Determine the overpressure from this vapor cloud explosion at a distance of 100 m from the blast using:
   a. the TNO multi-energy method
   b. the Baker–Strehlow method

Solution a: The heat of combustion of a stoichiometric hydrocarbon-air mixture is approximately 3.5 MJ/m^3 and, by multiplying by the confined volume, the resulting total energy is (2094 m^3)(3.5 MJ/m^3) = 7329 MJ. To apply the TNO multi-energy method a blast strength of 7 was chosen. The Sachs scaled distance is determined using Equation (2-31). The result is

$$\overline{R} = \frac{R}{(E/P_a)^{1/3}} = \frac{100 \text{ m}}{[(7329 \times 10^6 \text{ J})/(101,325 \text{ Pa})]^{1/3}} = 2.4$$

The curve labeled "7" on Figure 2.30 is then used to determine the scaled overpressure value of about 0.13. The resulting side-on overpressure is determined from Equation (2-32)

$$P_s = \Delta P_s \cdot P_a = (0.13)(101,325 \text{ Pa}) = 13,170 \text{ Pa} = 1.9 \text{ psi}$$

Solution b: The Baker–Strehlow pressure curves apply to free air blasts. Since the vapor cloud for this example is at ground level, the energy of the cloud is doubled to account for the strong reflection of the blast wave. The resulting total explosion energy is thus 2 × 7329 MJ or approximately 14,600 MJ.

The next step is to determine the flame speed using Table 2.11. Because the vapor cloud is enclosed beneath the storage tank the flame can only expand in two directions. Therefore, confinement is 2D. Based on the description of the piles the obstacle density is chosen as medium. The fuel reactivity for propane is medium. The resulting flame speed from Table 2.11 is 0.66. The Sachs scaled distance is determined from Equation (2-31) using a distance of 100 m. The result is

$$\bar{R} = \frac{R}{(E/P_a)^{1/3}} = \frac{100 \text{ m}}{[(14,600 \times 10^6 \text{ J})/(101,325 \text{ Pa})]^{1/3}} = 1.91$$

The scaled pressure is found from Figure 2.31 and has a value of about 0.13. The resulting overpressure is (0.13)(101,325 Pa) = 13,200 Pa = 1.9 psi. The TNO multi-energy method produces the same result.

2.10. Runaway Reactions

Figure 2.33 shows a batch reactor system. An exothermic reaction occurs within the reactor. The reaction temperature is controlled by the cooling water flow

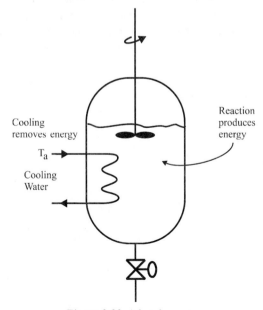

Figure 2.33. A batch reactor.

through the coils. The contents are well stirred, meaning that the temperature and species concentrations are uniform within the reactor. This is classified as a uniform reaction, as shown in Figure 2.1.

Assume that at some point the cooling water flow is lost. Since the reaction is exothermic, the loss of cooling will result in a higher temperature in the reactor. The higher temperature will increase the reaction rate resulting in a higher rate of heat generation. The result is a **runaway reaction**, where the reactor temperature increases exponentially. For very reactive chemicals, such as acrolein, hydroxylamine, and ethylene oxide, the temperature can increase several hundred degrees Celsius per minute.

The high temperature causes increased pressure within the reactor due to an increase in the vapor pressure of the liquid and/or generation of noncondensible gaseous products. It is also possible that the high temperatures could result in an additional decomposition reaction. The "decomp" might be more exothermic than the original, desired reaction. In either case, the pressure can rapidly exceed the burst pressure of the vessel, and result in an explosion.

Runaway reactions are possible whenever an exothermic reaction is encountered. This can occur within a reactor, storage vessel or even an open pool or container.

Table 2.14 shows the leading causes for vessel overpressure in the United States during the period 1980 to 1991. Clearly, the largest number of losses is due to runaway reactions, comprising 29% of the total losses.

Table 2.15 shows the common causes for runaway reactions based on a U.K. study. Loss of temperature control, including inadequate cooling or excessive heating, is the largest cause. This is followed by incorrect charging (wrong, too

TABLE 2.14
Causes of Vessel Overpressure Losses in the United States During 1980–1991 for Gross Losses Over $100,000 (Leung, 1997)

Cause	Number of Losses	Percent
Runaway Reaction	7	29
Plugged/No Relief	4	17
Scale-up	3	13
Unusual Procedure	3	13
Contamination	2	8
Inadequate Cooling	2	8
Instrument Failure	1	4
Unknown	2	8
Total	24	

TABLE 2.15
Common Causes for Runaway Based on U.K. Study (Leung, 1997)

Temperature control
  Inadequate cooling
  Excessive heating
	24%

Incorrect charging	18%

Reaction chemistry
  Unknown exothermal
  Unknown decomposition
	16%

Contamination	14%
Incorrect agitation	11%
Incorrect batch control	10%

much, or too little material), reaction chemistry (unknown exothermic behavior, or unknown decomposition at a higher temperature), contamination of reactants or vessel, incorrect agitation (too little, too much, failed, or delayed), and incorrect batch control (incorrect sequencing of operations).

After a runaway reaction is initiated the high pressure and bursting of the reactor might occur within minutes, or it might take hours, depending on the particular reactive system, heat transfer, etc. With everything else being equal, larger reaction vessels will runaway faster than smaller vessels since the surface to volume ratio is smaller and heat losses to the surroundings are less.

A number of factors have been identified that can lead to runaway reaction, as shown in Table 2.16. When any two or more of these factors are present, there is a potential for a thermal runaway. Thus, for a loss of coolant accident involving a

TABLE 2.16
Some Factors That Can Lead to Reactor Runaway (AIChE, 1995a; Regenass, 1984)

When any two or more of these factors are present, there is a potential for a thermal runaway.

1. High heat release of intended reaction
2. High heat release of potential decomposition
3. High heat release of competing reactions
4. Accumulation of reactants or intermediates
5. Insufficient heat removal
6. Thermally hazardous materials involved
7. High temperature
8. Loss of solvent used to control temperature
9. Improper mixing

very exothermic reaction, the two factors in Table 2.16 are "insufficient heat removal" and "high heat release of intended reaction."

An example of a particular type of runaway is the "sleeping reaction." This occurs in semi-batch reactors and has resulted in a number of explosions in the chemical industry. The problem is caused by a low reactor temperature while a reactant is being added. The low temperature causes a lower reaction rate, resulting in an accumulation of reactant in the reactor vessel. The reactant concentration will increase beyond the normal concentration expected during addition. If the reactor temperature is then increased, a runaway may occur if the heat generation rate from the reaction exceeds the capacity of the heat removal system. For this case, the two conditions from Table 2.16 are the accumulation of reactants and insufficient heat removal.

Another example is a runaway reaction caused by inadequate mixing. Suppose two liquid materials with differing densities and miscibilities are reacted in a semi-batch reactor. The first material is added to the reactor and then the second material is added slowly to control the reaction energy generation. However, due to an operator error, the mixer is not started prior to the addition of the second material. This results in a stratification of the liquids in the reactor vessel, with one liquid layer on top of the other. If the mixer is started after appreciable material has been added, a runaway and explosion can occur. Even if the agitation is not started, the reaction at the liquid interface may cause the phases to mix suddenly leading to a runaway. The two factors from Table 2.16 are the accumulation of reactants and improper mixing.

2.10.1. Steady-State and Dynamic Reactor Behavior

An introduction to reactor system dynamics is required to understand the causes and behavior of runaway reactions. More detailed information is available elsewhere (Benuzzi and Zaldivar, 1991; AIChE, 1995b).

Consider a semi-batch reactor where one reactant is fed at a constant rate during the reaction. The temperature of the reactor is controlled by both the feed rate and heat transfer through the cooling coils. Differential equations can be written representing the time rate of change of both the mass and energy within the reactor vessel. The time derivatives are set to zero to solve for the steady state temperature and reactant concentration. The steady state temperature and reactant concentration equations are algebraically combined, resulting in the following algebraic equation,

$$\frac{(-\Delta H)mMR(T)}{m + MR(T)} = mC_{\mathrm{P}}(T - T_0) + (UA)(T - T_{\mathrm{w}}) \tag{2-34}$$

where $(-\Delta H)$ is the enthalpy change during reaction (energy/mass)

      m is the feed rate of reactant (mass/time)

M is the total reaction mass in the reactor (mass)

$R(T)$ is the Arrhenius reaction rate (time^{-1})

C_P is the heat capacity of the feed (energy/mass-deg)

T is the temperature of the reactor (deg)

T_0 is the feed temperature (deg)

(UA) is the heat transfer coefficient for the cooling coils
           (energy/deg-time)

T_w is the temperature of the cooling water in the cooling coils (deg)

The left hand side (LHS) of Equation (2-34) is defined as the energy generation by the chemical reaction. The right hand side (RHS) of Equation (2-34) is the energy removal. This includes the energy required to heat the incoming cold reactant stream and the energy removed by the cooling coils. A plot of both the energy generation and removal curves versus the reactor temperature will yield a steady state curve, as shown in Figure 2.34. The intersections of the two curves indicates the locations of the steady states.

The energy generation term is the standard Arrhenius form, increasing exponentially with reactor temperature and then flattening out due to reactant consumption, as shown by the LHS of Equation (2-34). The energy removal term is linear with reactor temperature, as shown by the RHS of Equation (2-34). Both curves can be plotted together, as shown in Figure 2.34a.

For a continuous reactor, with both continuous input and output flows, the heat generation and removal curves are similar in shape to the semi-batch case, shown in Figure 2.34a. For this case, energy is also removed by the flow of hot mass out of the reactor. This additional energy removal term has the net effect of reducing the slope of the energy removal line.

For a batch reactor, a steady state never occurs since the reactant concentration decreases until the reactant is all consumed. However, if the assumption is made that the concentration remains relatively constant over short time intervals,

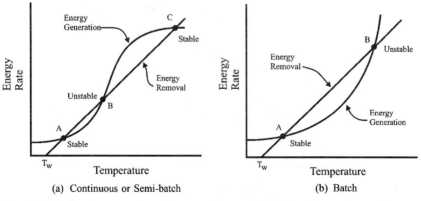

Figure 2.34. *Energy generation and removal curves. The intersections represent the steady states, some of which are unstable.*

then the results of Figure 2.34b are obtained. In this case the energy generation curve increases exponentially as the reactor temperature increases, but does not flatten at higher temperatures since the reactant concentration is assumed constant.

All three types of reactors, semi-batch, batch, and continuous, can be represented by curves similar to those shown in Figure 2.34.

The intersections of the heat generation and removal terms are the steady states for the reactor. Steady states A and C on Figure 2.34a are stable, whereas state B is unstable. States A and C are stable since an increase in reactor temperature will result in greater heat removal than generation, driving the system back to the steady state. State B is unstable since an increase in temperature will result in greater heat generation than heat removal, driving the system away from the steady state.

The stable steady states define the normal operating temperatures of the reactor. For the batch reactor, only one stable, operating temperature is observed at point A on Figure 2.34b. For the semi-batch and continuous reactors, two stable temperature states are possible, points A and C on Figure 2.34a. The high temperature state (point C) represents the temperature with high reactant conversion— most of the reactant is consumed. This is typically the normal and desired operating state of the reactor. The low temperature state (point A) is called the extinguished state, representing a state with low reactant consumption. This state is normally not desired due to low reaction conversion.

Figure 2.35 shows the effect on the steady states as the temperature of the cooling water, T_w, is increased and the heat transfer coefficient is assumed constant. This shifts the energy removal line to the right. Both the upper and lower steady state temperatures are increased as the cooling water temperature is increased. Note that a critical temperature is eventually reached, $T_{critical}$, where the

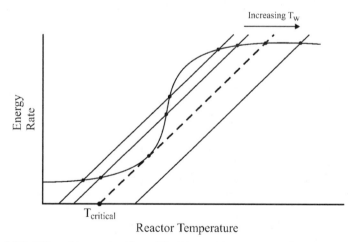

Figure 2.35. *Effect of increasing the cooling water temperature on the reactor steady state.*

heat removal curve is just tangent to the heat generation curve. Any increase in the cooling water temperature above $T_{critical}$ results in a single, high temperature steady state. The transition from one steady state to another might present a safety problem since the reactor experiences a rapid temperature and reactant conversion change.

Figure 2.36 shows the effect on the steady states as the heat transfer coefficient of the cooling coils is reduced and the cooling water temperature is assumed constant. In this case the slope of the energy removal line is decreased as the heat transfer coefficient decreases. Both the upper and lower steady state temperatures increase as the heat transfer coefficient decreases. Note that a critical condition exists here also where the energy removal curve is tangent to the energy generation curve—any further decrease in the heat transfer coefficient will result in a single, high temperature steady state. This transition from a low temperature to a high temperature steady state might present a safety problem due to the sudden increase in temperature.

If the cooling coil heat transfer coefficient is zero, that is, no energy removal through the coils, then the energy removal curve would be horizontal with a slope of zero. In reality, some heat transfer to the surroundings through the reactor wall is always present and the zero slope condition is not achieved.

An interesting situation occurs when cooling is lost to the reactor. This would occur due to an interruption in cooling water supply, accidental closing of a supply valve, plugging of a supply line, and so forth. This case can be represented by a sharp decrease in the heat transfer coefficient of the cooling coils, shown in Figure 2.36. Thus, if cooling is lost, the steady state temperature of the reactor will increase suddenly to a new, much larger value.

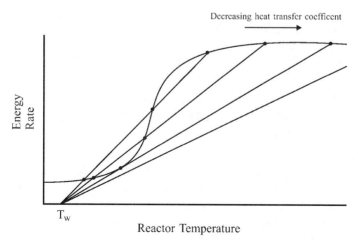

Figure 2.36. *Effect of decreasing the heat transfer coefficient of the cooling coil.*

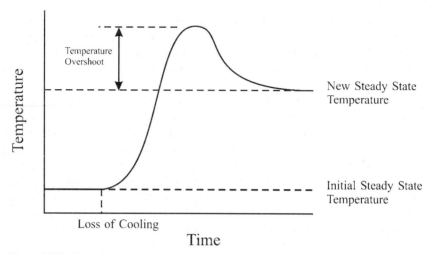

Figure 2.37. *Example temperature history for a loss of cooling incident. An overshoot or excursion in temperature is observed.*

The diagrams of Figures 2.34 through 2.36 present only the steady state cases and do not consider reactor dynamics. If the complete reactor dynamics are considered, the reactor will most likely make a high temperature excursion when a new, higher temperature steady state is defined, as shown in Figure 2.37. The reactor temperature will rapidly overshoot the final steady state value and then asymptotically approach that final steady state. This overshoot may cause a transient high temperature and pressure effect within the reactor resulting in possible reactor failure.

2.10.2. Experimental Characterization

Several commercial experimental devices are available to characterize the thermal behavior of reactions. These include:

1. APTAC™—Automatic Pressure Tracking Adiabatic Calorimeter
2. ARC™—Accelerating Rate Calorimeter
3. RSST™—Reactive System Screening Tool
4. VSP™—Vent Sizing Package

These devices are described in more detail elsewhere (AIChE, 1995a). An additional device, called a DSC for Differential Scanning Calorimeter, is also frequently used as an initial screening tool.

The primary purpose of these devices is to determine the adiabatic self-heat rate of the reaction system. The adiabatic self-heat rate is the maximum temperature rate of the material without heat losses. Once the self-heat rate is determined, then it is possible to design an adequate protection system. The experimental precision of the devices varies, but is typically on the order of 0.02 to 0.1°C/min.

These devices use a small sample, typically 5–30 ml, although the DSC uses milligram sized samples. The sample is loaded into the test cell at a specified composition. The device heats the sample using a variety of methods. One method is to heat to a fixed temperature and then wait while looking for sample self-heating. Another method is to heat the sample at a fixed temperature rate while observing the self-heat rate of the sample.

Figure 2.38 presents data obtained from the RSST™ for the reaction of methanol and acetic anhydride. The heat-up rate is 0.5°C/min and the test sample is pressurized with nitrogen at 300 psig to suppress liquid boiling and vapor stripping. The maximum temperature occurs at about 4900 seconds into the run with a value of 172°C. The slope of the temperature curve can be numerically determined and plotted versus −1000/T, as shown at the bottom of Figure 2.38. The straight line result over a large temperature range confirms a first-order reaction for this system.

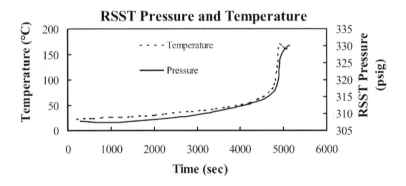

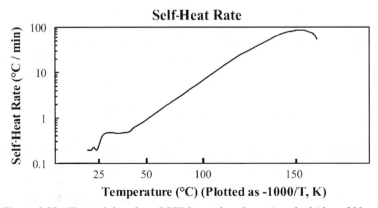

Figure 2.38. Thermal data from RSST for methanol–acetic anhydride at 300 psig.

Scaling of experimental results to a large scale reaction system represents a major problem. The idea is for the experimental device to mimic process conditions as closely as possible. However, for a large reactor, the surface to volume ratio is quite small. This is not true for a small sample. One method of characterizing this scaling is by the ϕ factor, defined as

$$\phi = 1 + \frac{m_r C_{P_r}}{m_s C_{P_s}}$$
(2-35)

where m_s is the mass of the reacting sample (mass)
  C_{P_s} is the specific heat capacity of the reacting sample (energy/mass-deg)
  m_r is the mass of the reactor (mass)
  C_{P_r} is the specific heat capacity of the reactor (energy/mass-deg)
  Physically, the ϕ factor represents the following

$$\phi = 1 + \frac{\text{Heat capacity of reactor vessel}}{\text{Heat capacity of the reactor contents}}$$
(2-36)

Clearly, for a very large reactor, $\phi \rightarrow 1$ since the sample heat capacity is much larger than the reactor vessel heat capacity. Typical process reactors have a ϕ factor of about 1.1. For a small sample typically found in a calorimeter, the ϕ factor will be much greater than 1 since the heat capacity of the test cell is comparable to the sample heat capacity. Ideally, the calorimeter should have a ϕ factor which is as close to 1 as possible in order to match process conditions. The ARC™ has a ϕ factor of about 1.5 to 3.0, while the VSP™ and APTAC™ approach 1.05.

A common experimental approach to achieve a low ϕ factor is to use a thin-walled test cell as shown in Figure 2.39. In order to prevent the test cell from rupturing due to reaction pressure, a pressure control system is used to ensure that the pressure outside the test cell is nearly identical to the pressure inside the test cell. This is the approach used in the VSP™ and APTAC™.

A more complete discussion of the experimental characterization of the thermal behavior of energetic materials is well beyond the scope of this book. This characterization is described in much greater detail elsewhere (ESCIS, 1993; AIChE, 1995a, 1995b).

2.11. Condensed Phase Explosions

A **condensed phase explosion** occurs when a solid or liquid material explodes directly from the bulk state. High explosives, such as TNT, are a common example of a condensed phase explosive.

There are many solid and liquid materials capable of undergoing very sudden and explosive decomposition or reaction due to exposure to, for example, shock,

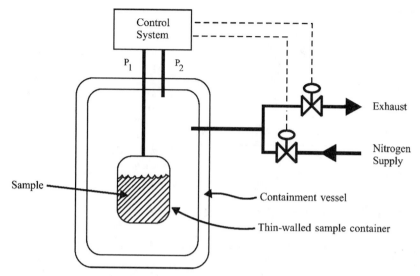

Figure 2.39. An experimental setup to achieve low ϕ factor. The control system balances the pressure inside and outside the sample container, preventing its rupture.

temperature, pressure, and contamination. Examples of these materials include many peroxide and nitrate materials.

The 1947 explosion in Texas City is an excellent example of a condensed phase explosion due to the sudden decomposition of ammonium nitrate contained in the hull of the ship *Grandcamp* (Lees, 1996). The ammonium nitrate was exposed to fire in the ship's hold for some time prior to the explosion. This explosion resulted in 552 fatalities and almost complete destruction of everything for miles from the blast.

The equilibrium state for a system is defined as the state with the minimum Gibbs energy. Many chemicals, however, demonstrate a metastable state, shown in Figure 2.40 as the high level trough denoted as point A. If energy is provided that exceeds the trough or activation energy, then a reaction will occur moving the material to the new equilibrium state, denoted as point B. The reaction may occur very suddenly, with the potential release of large volumes of gas or vapor. An explosion may result, with damage due to the blast wave, thermal energy or projectiles.

The experimental characterization of these materials is beyond the scope of this book. Additional information is provided elsewhere (AIChE, 1995a).

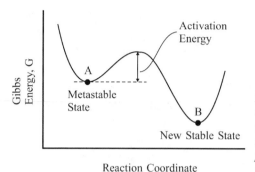

Figure 2.40. A Gibbs energy plot demonstrating a metastable equilibrium state, point A. The system will move to a new equilibrium state, point B, if the activation energy is provided.

2.12. Fireballs, Pool, Flash, and Jet Fires

The primary effects of fireballs, pool, flash, and jet fires are due to thermal radiation and direct flame contact.

A **fireball** results from a burning fuel-air cloud. The inner core of the cloud consists almost entirely of fuel, whereas the outer layer (where ignition first occurs) consists of a flammable fuel-air mixture. As the buoyancy forces of the hot combustion gases increase, the burning cloud tends to rise, expand, and assume a spherical shape. Most BLEVEs involving flammable or combustible liquids result in fireballs since the liquid first flashes into vapor and then burns on the outer surface as the fuel mixes with air. Fireballs are typically of short duration, but very high thermal radiation flux. A fireball resulting from a BLEVE may be up to several hundred feet in diameter.

Pool fires are the result of surface burning of flammable or combustible liquid. These fires tend to be localized in effect and are mainly of concern for establishing the potential for domino effects and employee safety zones, rather than for assessing community risk.

A pool fire begins typically with the release of flammable or combustible material from process equipment. If the material is a liquid, stored at a temperature below its normal boiling point, the liquid will collect in a pool. The geometry of the pool is dictated by the surroundings, for example, diking, but an unconstrained pool in an open, flat area is possible. An unconstrained pool is also possible if the liquid quantity spilled is inadequate to completely fill a diked area or overflows the dike. If the liquid is stored under pressure above its normal boiling point, then a fraction of the liquid will flash into vapor, with a portion of the unflashed liquid remaining to form a pool in the vicinity of the release.

Important questions for pool fires are: Where can the liquid go? How far can it travel? This is particularly true of spills from railroad tank cars or roadway trucks which might be off-site or at a location without adequate spill containment or emergency response.

Pool fire modeling is well developed (Mudan and Croce, 1988; AIChE, 1999a). The determination of the thermal effects depends on the type of fuel, the geometry of the pool, the duration of the fire, flame tilt (due to wind or buoyancy), the location of the radiation receptor with respect to the fire, and the thermal behavior of the receiver, to name a few. These effects are treated using separate, but interlinked models.

For large hydrocarbon pool fires, especially those involving high molecular weight fuels, large amounts of soot are usually generated, obscuring the radiating flame from the surroundings, and absorbing much of the radiation. Flame turbulence and wind causes the smoke layer to open up occasionally exposing the hot flame and increasing the radiative flux temporarily.

Typical surface emitted radiative flux values for LPG and LNG pool fires are about $250 \, kW/m^2$ ($79,000 \, Btu/hr-ft^2$). Upper values for other hydrocarbon fires lie in the range $110–170 \, kW/m^2$ ($35,000–54,000 \, Btu/hr-ft^2$), but smoke obscuration often reduces this to $20–60 \, kW/m^2$ ($6300–19,000 \, Btu/hr-ft^2$). For comparison, solar radiation intensity on a clear, hot summer day is about $1 \, kW/m^2$ ($320 \, Btu/hr-ft^2$).

A **flash fire** is the nonexplosive combustion of a vapor cloud resulting from a release of flammable or combustible material into the open air. Premixing with air of some part of the vapor cloud is required for a flash fire. A flash fire is nonexplosive since the flame speed has not accelerated sufficiently to produce damaging overpressures. Flash fires do not create a blast; however, a delayed ignition of a flammable vapor which has accumulated in a congested area, such as a process plant, may result in a vapor cloud explosion.

The literature provides little information on the effects of thermal radiation from flash fires. Furthermore, flash combustion of a vapor cloud normally lasts no more than a few seconds. Therefore, the total intercepted radiation by an object near a flash fire is substantially lower than in the case of a pool fire.

Jet fires result from the combustion of a material as it is being released from a pressurized process unit. The velocity of the released material contributes significantly to the behavior of the jet fire. First, the release velocity must be high enough to entrain a considerable quantity of air into the jet. Second, the flame will stabilize at a point in the jet where the turbulent flame speed is the same as the local mixed gas velocity. As the release velocity is increased the air entrainment increases, the concentration of fuel in the jet decreases, and the flame point will move downstream from the release point. Conditions are eventually reached where the flame point is so far downstream that the fuel concentration is below the flammability limit and flame blow-off occurs.

Flame stabilization of a jet could also be achieved by an obstruction or obstacle in the jet's path.

Jet fire modeling is not as well-developed as pool fire modeling. A review of available models is presented elsewhere (AIChE, 1999a).

2.13. Explosion Effects

Explosions can injure or kill people and damage equipment by several mechanisms, including thermal exposure, overpressure, and flying debris or missiles.

2.13.1. Thermal Exposure

Under specific conditions and, depending on the type of material, the release and ignition of a combustible material may result in a pool fire, flash fire, jet fire or a fireball. The energy conveyed to the surroundings due to thermal radiation and convection are also dependent on the release and the release environment. The main hazard associated with large fires is thermal radiation. Thermal radiation effects may be difficult to estimate since the radiation varies with the fourth power of temperature and are thus subject to large error.

Reviews of thermal radiation effects are available (Buettner, 1957; Stoll and Green, 1958). Table 2.17 provides human thermal exposure data based on the time to reach the pain threshold. The thermal radiation exposure for a hot summer day is provided for perspective, although great variability in this value is expected based on location and time of day.

Mudan (1984) has reviewed the work of Eisenberg et al. (1975) on burns and fatalities from thermal radiation. Figure 2.41 shows the injury and fatality levels for thermal radiations. Schubach (1995) provides a review of thermal radiation and suggests an intensity of 4.7 kW/m^2 to represent the threshold injury. This radiation would cause injury after 30 seconds of exposure possibly affecting persons unable to seek shelter. Further details in thermal exposure is provided by Hymes (1983).

TABLE 2.17
Human Thermal Exposure Data Based on Time to Pain Threshold (API 521, 1999)

Radiation intensity		
(Btu/hr-ft^2)	(kW/m^2)	Time to pain threshold (sec)
320	1.00	Hot summer day
500	1.74	60
740	2.33	40
920	2.90	30
1500	4.73	16
2200	6.94	9
3000	9.46	6
3700	11.67	4
6300	19.87	2

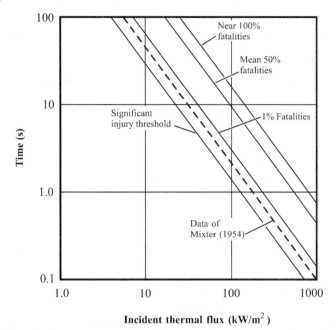

Figure 2.41. *Injury and fatalities for thermal radiations (Mudan, 1984).*

The effect of thermal radiation on process equipment and building materials depends on the duration and nature of the exposure. Thermal radiation may be reflected, partially absorbed or pass through to other materials. It is the radiation absorbed by a material that produces heat resulting in damage to the material. The absorbed radiation depends on the color and nature of the material. The thermal radiation passes through thin materials while most of it is reflected by white materials. Black materials, however, absorb a much larger proportion of the incident radiation. Table 2.18 shows the effects of thermal radiation on personnel and materials. Wood-based construction materials will fail due to combustion, whereas steel fails due to thermal lowering of the yield stress. Steel structures heated to 500–600°C will fail rapidly under normal loads as, for example, is proposed for the collapse of the World Trade Center towers on September 11, 2001.

2.13.2. Overpressure Exposure

Due to the geometry and size of the human body, a blast or shock wave results in initial compression followed by a drag force often resulting in body translation with possible subsequent impact. Fast compression and decompression of the human body results in transmission of pressure waves that damage air-containing

TABLE 2.18
Effects of Thermal Radiation on Personnel and Materials (World Bank, 1985)

Radiation intensity (kW/m^2)	Observed effect
37.5	Damage to process equipment
25.0	Minimum energy required to ignite wood at indefinitely long exposures unpiloted
12.5	Minimum energy required for piloted ignition of wood, melting plastic tubing
9.5	Pain threshold reached after 8 sec; second degree burns after 20 sec
4.0	Sufficient to cause pain to personnel if unable to reach cover within 20 sec; however blistering of the skin (second degree burns) is likely; 0% lethality
1.6	Will cause no discomfort for long exposure

organs (lungs, ears, sinuses, and orbital bones near the sinus cavities) and the interfaces between tissues of varying density.

The lungs are very susceptible to overpressure, which results in lung hemorrhage. Hemorrhage may lead to death within minutes due to obstruction of the airways by fluid. Lung damage can occur at overpressures as low as 15 psig (103 kPa).

The extent of ear drum damage is a function of age, peak overpressure, and duration of the exposure. Damage may occur at overpressures from 5 to 45 psig (34.5 to 310 kPa), with an expected 50% probability of rupture at about 25 psig (172 kPa).

Human injury from blast debris depends on many factors. These include velocity, angle of impact, size, density, mass and the portion of the body involved. Glass is one of the more common and potentially more injurious blast debris—it shatters easily and generates large quantities of glass shard missiles.

Human injury also results from the accelerative component of blast wave translation, particularly by abrupt deceleration from impact against a solid object. An impact velocity of 10 ft/sec (3.0 m/s) is unlikely to produce significant injuries. Impacts of 10 to 20 ft/sec (3.0 to 6.0 m/s) will result in some fatalities. Above 20 ft/sec (6 m/s) the likelihood of fatal injury increases sharply with increasing velocity.

Damage to houses and other unreinforced structures can occur at overpressures as low as 0.4 psig (2.7 kPa), with complete demolition of the structure around 3 psig (28 kPa). These overpressures are well below the pressures for significant human injury from the direct effects of the blast or shock wave. However, any humans within these structures are likely to be seriously injured or killed. To adequately address the risk to humans inside of buildings, the analysis must address the potential for the explosion event, the potential for fatalities due to building damage, and the amount of time people spend in the building (AIChE, 1996b).

When a high explosive, such as TNT, detonates, adjacent objects are broken into a large number of small fragments with high velocity and chunky shape (AIChE, 1994). In contrast, a BLEVE or other deflagrative explosion produces only a few fragments, varying in size, shape and initial velocities. Fragments may travel long distances because large, half-vessel fragments can "rocket" and disk-shaped fragments can "frisbee." Baker et al. (1988), Brown (1985, 1986), and AIChE/CCPS (AIChE, 1994; AIChE, 1999a) provide equations for prediction of projectile effects. These projectiles can cause human injury and damage to nearby equipment and communities. In addition, damage to equipment in adjacent processing units from these projectiles can cause subsequent release of flammable and/or toxic materials, resulting in an expansion of the accident.

There are two main considerations in evaluating the effect of a blast or shock wave on equipment or structures. The first is the loading, that is, the forces from the blast or shock wave acting on the structure. The second is the response of the structure to the loading. The response of the structure results in damage which could include total destruction of the structure; an intermediate response such as deflected steel frames, collapsed roofs, dished walls, cracked masonry and broken windows; or no damage at all.

Figure 2.42 shows the interaction of a shock wave with a rigid structure. In the first sequence the shock wave is about to interact with the structure. In the second sequence the shock wave has impacted the front of the structure resulting in a reflected shock wave and a resulting overpressure. Weak waves will result in reflected overpressures just greater than double the incident side-on overpressure, while strong shock waves can result in overpressures as high as eight times the side-on overpressure. The reflected wave travels left due to its interaction with the structure—as it does a rarefaction front moves downward along the structure. This wave continues to the end of the structure where it diffracts around the edge as in the third sequence. The vortices produced result in areas of high and low pressure, causing variable forces on the structure. In the fourth sequence the structure experiences the transient blast wind which exerts drag forces.

The blast **impulse** is defined as the change in momentum and has dimensions of force-time product. For a blast wave, the area under the pressure—time curve is the impulse per unit of projected area (Kinney and Graham, 1985)

Numerous factors determine the response of a structure to a blast wave. These include strength and mass of the structure, design and ductility of the materials of construction. The damage also depends on how quickly the structure responds to the blast wave. Small, rigid structures will respond much more quickly to a blast wave than large, flimsy structures. A measure of this response is called the natural period of vibration. If the duration of the blast wave is long compared to the natural

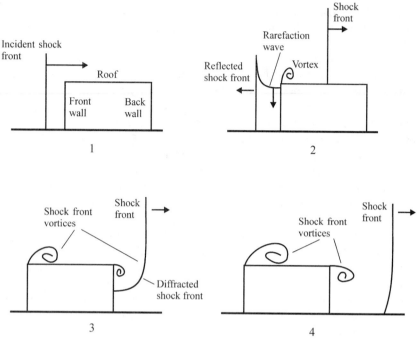

Figure 2.42. Interaction of a blast wave with a rigid structure (ASCE, 1997).

period of vibration of the structure, then the loading can be approached as suddenly applied, constant pressure and the blast damage is determined from the peak overpressure. If the duration of the blast wave is short compared to the natural period of vibration of the structure, then the damage is determined from the impulse. This is shown in Figure 2.43 which is a pressure impulse diagram for a structural component with a single degree of freedom exposed to an ideal shock wave. The line on Figure 2.43 represents a line of constant damage. The impulsive and pressure regions are clearly shown. The dynamic region is where the damage depends on both the overpressure and the impulse.

A significant percentage of the accident scenarios for the chemical processing industry are in the dynamic region. Thus, great caution must be exercised when using damage tables that rely on peak overpressure only, such as Tables 2.9 and 2.10. However, for very preliminary damage estimates, the overpressure alone can be used. Oswald (2001) provides detailed pressure-impulse diagrams for many types of construction commonly found in chemical plants. Computation fluid dynamic (CFD) codes can also be used to more detailed estimates.

Many risk analysts use 3 psi (20.7 kPa) as a conservative end point for quantitative risk calculations. That is, for overpressures greater than 3 psi (20.7 kPa), the structures are considered completely damaged and all of the humans in the struc-

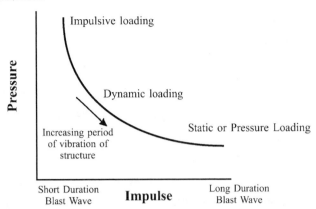

Figure 2.43. *Pressure–impulse diagram for a single degree of freedom (Baker and Cox, 1988; Lees, 1996).*

tures result in fatalities. This assumes that humans are within the damaged buildings and fatalities are almost certain due to the collapsing structure (AIChE, 1999a, 2000). Oswald and Baker (1999) provide considerably more detail on fatalities due to building collapse.

Structures can be designed for blast resistance—this subject is beyond the scope of this book. More detailed information can be found elsewhere (ASCE, 1997).

2.14. Ignition Sources

The fire triangle in Figure 2.2 shows that an ignition source is one of the requirements for a fire or explosion. As was discussed in Section 2.1, ignition sources are too numerous and elusive to depend on elimination of ignition sources as the sole defense against fires and explosions. The current practice to prevent fires and explosions is to focus strongly on preventing flammable mixtures while at the same time striving to eliminate ignition sources. An understanding of ignition sources is essential to this practice.

Table 2.19 shows the results of a survey performed by Factory Mutual Engineering based on over 25,000 fires. The results show that electrical, smoking, friction, hot surfaces, open flames and sparks account for a large percentage of the ignitions. The hot surface of an exposed light bulb, electric motor, or piece of process equipment can provide a suitable ignition source. A more detailed discussion of ignition sources is provided elsewhere (Bond, 1991). Most of the ignition sources shown in Table 2.19 can be controlled using proper management systems (such as a hot-work permit system) and by proper equipment and process design (such as using electrically rated equipment).

TABLE 2.19
Ignition Sources of Major Fires
(National Safety Council, 1974)

Electrical (wiring of motors)	23%
Smoking	18%
Friction (bearings or broken parts)	10%
Overheated materials (abnormally high temperatures)	8%
Hot surfaces (heat from boilers, lamps, etc.)	7%
Burner flames (improper use of torches, etc.)	7%
Combustion sparks (sparks and embers)	5%
Spontaneous ignition (rubbish, etc.)	4%
Cutting and welding (sparks, arcs, heat, etc.)	4%
Exposure (fires jumping into new areas)	3%
Incendiarism (fires maliciously set)	3%
Mechanical sparks (grinders, crushers, etc.)	2%
Molten substances (hot spills)	2%
Chemical action (process not in control)	1%
Static sparks (release of accumulated energy)	1%
Lightning (where lightning roods are not used)	1%
Miscellaneous	1%

When flammable or combustible materials are present the electrical equipment must be specially designed to reduce the probability of causing an ignition. The area where the flammable or combustible materials are used is called a "classified" area. The electrical equipment is designated for use in classified areas. The specific classification and electrical equipment required is a function of the physical properties of the flammable or combustible material and the probability of having a combustible mixture in the area. More detailed information is provided in Section 3.13.

The type of fire that occurs depends on when and how the flammable material is ignited. For materials released below their autoignition temperature an ignition source is required. Each release has a finite probability of ignition. For liquid releases, the ignition can occur via the vapor cloud (for flashing or evaporating liquids), with the flame traveling upwind via the vapor to ignite the liquid pool. For liquids stored below their normal boiling point without flashing, the ignition can still occur via the flammable vapor from the evaporating liquid. Both of these cases may result in an initial flash fire due to burning of vapors—this may cause initial thermal hazards. If the release of flammable material continues after ignition, then a jet fire is also likely. If the ignition occurs at the very beginning of the release, then inadequate time is available for the liquid to form a pool and a jet fire will usually result.

2.14.1. Static Electricity

Static electricity is a serious problem in the chemical industry—considerable understanding and effort is required to control it. Static electricity, however, is not easily understood and requires a fundamental understanding of its behavior to control it properly. More detailed references on the subject are available (Pratt, 1997; Britton, 1999; NFPA 77, 2000; Crowl and Louvar, 2002).

In order to understand static electricity, the engineer must understand (Louvar, Maurer et al., 1994):

- How charge is generated
- How charge is accumulated on objects
- How charges are discharged
- How the discharged energy is related to the minimum ignition energy (MIE) of the material.

Every object has an inherent electric field as a result of the electrons in the material. This includes conductive as well as nonconductive materials. The electric field around isolated objects is different than the electric field around two objects that are physically connected. When two connected objects are suddenly separated, the electric field changes and the electrons must redistribute themselves to form a new electric field. If both objects are conductive, the electrons are able to move rapidly through and between the objects as they are being separated. The net result after separation is two objects at identical voltages. However, if one or more of the objects is nonconductive, the electrons are unable to move rapidly as the objects are being separated. The final result after separation is two objects with opposite, but equal charges. Separation is one mechanism for static charge generation.

Separation mechanisms for static charge generation include a variety of physical transport and processing mechanisms, such as pouring, flowing, grinding, sliding, pneumatic transport, sieving of powders, and atomization of liquid.

The microscopic separation of the liquid from any interface (solid–liquid, gas–liquid, liquid–liquid) generates static electricity. The charge carried away by the flowing liquid is called a **streaming current**, and can result in substantial charge generation. The most common example is the flow of liquid through a pipe.

Another mechanism for charge generation is by **induction**. If conductive object A is brought close to charged object B, the electrons in object A are attracted to one end and are repelled at the other end. This results in opposite charges on opposite ends of object A. If one end of object A is discharged to ground (perhaps via a spark), some of the electrons are lost. If object B is then removed, a charge is left on object A.

Another mechanism for charge generation is by **transport**. This occurs when charged liquid droplets or solid particles settle on an object, resulting in a transfer of charge to the object.

Two laboratory examples of static electricity generation are shown in Figure 2.44. In Figure 2.44a dry cellulose powder is poured out of a beaker with free fall into an insulated metal container. A special voltmeter, called an electrometer, is

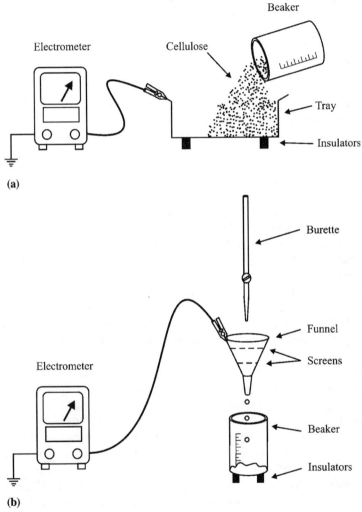

Figure 2.44. (a) Charge accumulation by pouring of cellulose powder. (b) Charge accumulation by xylene flow.

connected to the metal container. This voltmeter has a very high input impedance (about 10^{12} ohms) to prevent the static charge from leaking to ground through the meter. The experiment generates 36 volts per gram of cellulose powder. The rapid dumping of much larger masses of powder into a process tank would result in very high charge generation and voltage, leading to a possible discharge and dust explosion.

In the second experiment, shown in Figure 2.44b, xylene is metered out of a burette through a metal funnel with a screen, into a beaker. The screen is essential in this case to cause adequate separation—these types of screens are typically used to screen out debris in the liquid. The electrometer records the voltage accumulation on the funnel. The experiment results in the generation of about 30 volts per 100 ml of xylene.

Charge accumulation results from the competing processes of charge generation and charge dissipation. If charge generation is larger than charge dissipation, then a net charge accumulates on an object.

The voltage on a conductor is defined as the charge (in Coulombs) divided by the capacitance (in Farads), or $V = Q/C$. The total energy accumulated on a conductor is calculated using the following equations from electrical theory,

$$J = \frac{Q^2}{2C} = \frac{1}{2}CV^2 = \frac{1}{2}QV \qquad (2\text{-}37)$$

where J is the total energy accumulated (Joules)

C is the capacitance (Farads)

Q is the charge (Coulombs)

V is the voltage (Volts)

For unit conversions, a Farad is a Coulomb/Volt and a Joule is a Coulomb × Volt.

Equations are available to estimate streaming currents, capacitances of various shapes and for estimating charge dissipation through liquid pools (Louvar, Maurer et al., 1994; Britton, 1999; Crowl and Louvar, 2002).

For the experiments shown in Figures 2.44a and 2.44b, the total energy accumulated on the metal tray or beaker is rather small due to the low capacitance of these items. Since $V = Q/C$, if the capacitance, C, is low, a small charge, Q, will result in a large voltage, V.

A case history will demonstrate how a change in the procedure resulted in significant charge accumulation and ignition of a flammable vapor. Two workers were severely burned when a small explosion and fire occurred during the charging of powder into an open vessel containing toluene. For many years the powder charging operation was done manually using 50-lb sacks. An efficiency analysis concluded that the process throughput could be greatly improved by replacing the 50-lb sacks with a flexible intermediate bulk container (FIBC) holding 1 ton of powder. The FIBC would be hoisted above the vessel by a crane and the necessary powder dumped in a matter of minutes. Two weeks after the installation of the FIBC, the explosion occurred. The accident was attributed to ignition of the flammable vapors by charge generation and subsequent sparking due to the powder dumping. Previously, using the 50-lb sacks, the powder dumping occurred slow enough that the charge was able to dissipate during the powder charging operation. Dumping of powders into vessels with flammable atmospheres should be avoided, unless suitable explosion protection is provided (NFPA 68, 1994, NFPA 68 2002 #82; NFPA 69, 1997).

Several types of discharges are possible with static electricity (Louvar, Maurer et al., 1994; Britton, 1999). A **spark** occurs when charge moves through the air between two conductors, as shown in Figure 2.45. A spark is well defined and has a sharp, needle like appearance. The charge exits and enters both conductors from a single point. Energy levels for sparks can be as high as 1000 mJ, more than enough to ignite both combustible dusts and vapors. A **brush discharge** occurs between a conductor and a nonconductor, as shown in Figure 2.46. In this case the discharge exits from the conductor at a fixed point and has a fuzzy appear-

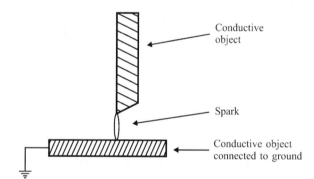

Figure 2.45. A spark discharge between two conductive objects.

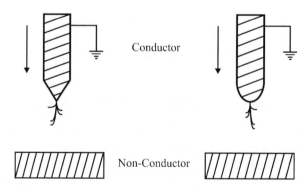

Figure 2.46. A brush discharge between a conductor and a nonconductor.

ance towards the nonconductor as the charge is absorbed across a part of its surface. Brush discharges have total energy levels up to about 4 mJ, enough to ignite most flammable vapors. Ignition of dusts by brush discharge has never been demonstrated experimentally (Britton, 1999). This is likely due to the low energy density in a brush discharge (although the total energy may exceed the MIE for some dusts). A **corona discharge** is a special case of a brush discharge, occurring between a charged nonconductor and a conductor with a small radius of curvature, for example, a conductor with a pointed shape. These discharges have limited energy and are only capable of igniting the most sensitive gases, for example, hydrogen. A **propagating brush discharge** can occur between a grounded conductor and a charged insulator which is backed by a conductor, as shown in Figure 2.47. This situation occurs with, for instance, metal drums lined with nonconductive plastic. These discharges are very impressive and will ignite flammable dusts and vapors, with energy levels on the order of several thousand millijoules. Experiments have shown that propagating brush discharges are not possible if the breakdown voltage of the insulator is 4 kV or less. A **conical pile discharge** (also called a **bulking brush discharge**) is a form of brush discharge that occurs at the surface of a pile of powder. The necessary conditions for this discharge are: a powder with high resistivity (greater than 10^{10} ohm-m), a significant portion of the particles with diameters greater than 1 mm, a high charge-to-mass ratio, and a filling rate above about 0.5 kg/s. These discharges have energies up to several hundred mJ and can ignite gases and dusts.

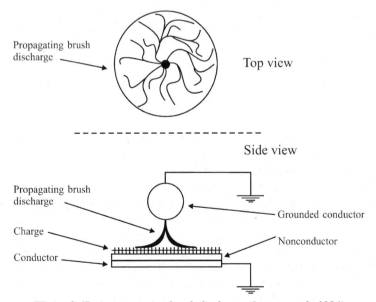

Figure 2.47. A propagating brush discharge (Louvar et al., 1994).

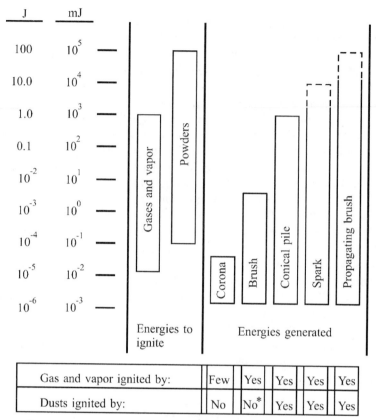

*Brush discharge energies are above the MIE for a number of dusts, but direct ignition of these dusts has not been demonstrated.

Figure 2.48. *Summary of discharge types (Louvar et al., 1994).*

Figure 2.48 summarizes the various types of discharges, their energies, and the types of materials they can ignite. Almost all charge accumulations are well above the MIE of flammable vapors and powders.

Figure 2.49 shows how charge accumulation in a glass-lined vessel can result in an explosion. Flammable liquid is splash filled into the vessel. Due to splashing and mixing with air, a flammable atmosphere is present in the head space of the vessel. A streaming current is generated as the liquid flows through the pipe prior to discharge into the vessel. Thus, the liquid dropping into the vessel has a net charge, which accumulates in the liquid. The glass lining of the vessel is nonconductive so the charge is unable to dissipate. Eventually the liquid level rises and approaches a grounded metal thermocouple. A spark or brush discharge may

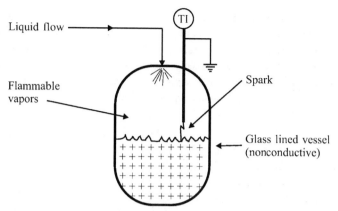

Figure 2.49. *Charge accumulation in a glass-lined vessel due to liquid free fall. A spark occurs between the liquid and the grounded thermowell.*

result between the liquid and the thermocouple depending on whether the liquid is conductive or nonconductive. An explosion may result.

Another possible charge accumulation mechanism is shown in Figure 2.50. In this case liquid is splash filled into the center of a very large storage vessel. The vessel is metal, conductive and grounded. As in the previous case, the liquid dropping into the vessel is charged due to the streaming current in the pipe. Charge may accumulate in the center of the liquid pool depending on relative magnitudes of the charge generation rate and the charge dissipation through the liquid to the grounded vessel. The charge dissipation through the liquid is a function of the

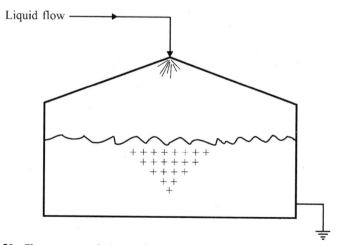

Figure 2.50. *Charge accumulation on the center of a large pool of liquid. The charge dissipates slowly through the liquid to the grounded vessel.*

distance to the grounded vessel wall and the resistivity of the liquid. If a charge accumulates in the center of the liquid pool, a discharge might occur between the liquid and a grounded metal structural object within the vessel, resulting in a fire or explosion. The **charge relaxation time,** τ is defined as the time for 37% (i.e., $1/e$) of the charge in the pool to dissipate once the flow of liquid into the tank has been stopped. The relaxation time is defined by the equation

$$Q = Q_0 e^{-t/\tau}$$

(2-38)

where Q is the charge at time t
     Q_0 is the charge at $t = 0$
     t is the time (time)
     τ is the relaxation time (time)
    If $t = \tau$ then from Equation (2-38), $Q/Q_0 = 0.37$.

3

PREVENTION AND MITIGATION OF EXPLOSIONS

This chapter will show how the fundamentals of combustion and explosion are used to prevent and/or mitigate an explosion.

In general, a safety program should have the following four components:

1. Identification and characterization of the major hazards in the process.
2. Application of inherently safer design concepts to eliminate or minimize the hazards.
3. Control of the hazards to prevent an incident.
4. Mitigation of the impacts of an incident, should it occur.

Item 1 is discussed in detail elsewhere (AIChE, 1992a). Item 2 is discussed in Section 3.2, with much more additional information provided elsewhere (Bollinger, Clark et al., 1996). Items 3 and 4 will be discussed generally, with additional references provided in Section 3.1.

3.1. Additional References

Table 3.1 provides detailed references on fires and explosions. These references are recommended for readers who want additional details on managing the particular hazards shown. Across the top of Table 3.1 are the various types of explosions. The left-hand column on Table 3.1 provides specific topics related to explosions. The shaded table entries indicates that no reference is available or provided. The references shown in Table 3.1 are not definitive—they are provided only to help the reader find a starting point for more information.

3.2. Inherently Safer Design

A chemical process is defined as inherently safer if "it reduces or eliminates the hazards associated with materials and operations used in the process and this reduction or elimination is permanent and inseparable" (Bollinger, Clark et al.,

TABLE 3.1
Detailed References for Various Types of Explosions

TOPIC:	Gases/ Vapors	Liquids	Aerosols and Mists	Dusts	Hybrid Mixtures	Physical Explosions	Vapor Clouds	Runaway Reactions	Condensed Phase Explosions	Pool, Flash and Jet Fires
Inherently Safer Design:	Kletz, 1991; Bollinger, et al., 1996; AICHE, 1993									
Inerting/Purging:	NFPA 69, 1997; Crowl and Louvar, 1990; 2002; AICHE, 1995c; AICHE, 1993									
Grounding and Bonding:	AICHE, 1993; AICHE, 1995c; AICHE, 2000; Crowl and Louvar, 1990; 2002; NFPA 77, 2000; NFPA 654, 2000									
Ventilation:	AICHE, 1995c; NFPA 30, 2000; NFPA 33, 2000; NFPA 86, 1999; NFPA 654, 2000									
Sprinkler Systems:	NFPA 15, 2001; NFPA 58, 2001; NFPA 59, 2001; NFPA 13, 1999; NFPA 230, 1999; API 2030, 1998						Same			
Charging and drumming:		Crowl and Louvar, 1990; 2002		Eckoff, 1997;						
Electrical Classification:	AICHE, 1995c; NFPA 30, 2000; NFPA 70, 2002; NFPA 496, 1998; NFPA 497, 1997; NFPA 499, 1997; NFPA 654, 2000; API 540, 1999			NFPA 69, 1997			Same		Same	
Explosion Prevention	NFPA 69, 1997			NFPA 69, 1997						
Explosion Venting:	AICHE, 1993; NFPA 68, 1998; NFPA 69, 1997			NFPA 68, 1998; Bartknecht, 1981	NFPA 68, 1998					
Fireproofing:	AICHE, 1993; NFPA 58, 2001; API 2218, 1999						Same			Same
Drainage:	CCPS, 1993; NFPA 30, 2000									Same

Category				
Mechanical Integrity:	Sanders, 1999		Sanders, 1999; AICHE, 1998; Fisher et al., 1992	Same
Combustibles concentration control:	NFPA 69, 1997			
Ignition Control:	Britton, 1999; Pratt, 1997; Jones and King, 1991; API 2003, 1998; NFPA 77, 2000	Same		Same
Management of Change:	Sanders, 1999; Kletz, 1991			
Plant layout:	API 752, 1995; AICHE, 1994; AICHE, 1993			
Pressure relief:	API 520, 2000; API 521, 1999; AICHE, 1993		AICHE, 1993; AICHE, 1998; Crowl and Louvar, 1990; 2002	
Explosion suppression systems:	NFPA 69, 1997	NFPA 69, 1997		
Pressure Containment:	AICHE, 1993; 1996b		AICHE, 1998; Fisher, et al., 1992	
Explosion Consequences:	AICHE, 1999a; Crowl and Louvar, 1990, 2002; Baker and Cox, 1988			
Blast resistant buildings:	ASCE, 1997; NFPA 654, 2000; AICHE, 1996b			

1996). A hazard is defined as a physical or chemical characteristic that has the potential for causing harm to people, the environment, or property. This means that the hazard is intrinsic to the material, or to its conditions of storage or use. These hazards cannot be changed—they are basic properties of the materials and the conditions of usage. The inherently safer approach is to reduce the hazard by reducing the quantity of hazardous material or energy, or by completely eliminating the hazardous agent.

One traditional risk management approach is to control the hazard by providing layers of protection between it and the people, property and surrounding environment to be protected. These layers of protection may include operator supervision, control systems, alarms, interlocks, physical protection devices, barriers and emergency response systems. This approach can be highly effective, but does have significant disadvantages. This includes cost, failure of the protection layers, and impacts realized through another, unanticipated failure route or mechanism.

For these reasons, the inherently safer approach should be an essential aspect of any safety program. If the hazards can be eliminated or reduced, the extensive layers of protection to control those hazards may not be required.

The four strategies for inherently safer design are:

Minimize: To reduce the quantity of material or energy contained in the manufacturing process or plant. This includes reducing inventories of hazardous chemicals and reducing the size of the processing equipment.

Substitute: The replacement of the hazardous material or process with an alternative which reduces or eliminates the hazard. This includes selecting a solvent that is less flammable, or less toxic, or using reaction chemistry that uses less hazardous materials.

Moderate: Using materials under less hazardous conditions. This includes using lower temperature or pressures, or selecting reaction chemistry which operates at less severe conditions.

Simplify: Designing to eliminate unnecessary complexity, thus reducing the opportunities for error and mis-operation. This includes simplifying control panel layout, labeling piping and equipment, and selecting a chemistry that involves less processing steps.

Inherently safer design is very important for preventing fires and explosions. Several questions come to mind: Why are we using that flammable solvent—can we use one that is not flammable? Why are we operating at this high pressure—can we reduce the pressure? Do we really need such a large inventory of flammable solvent? Can we use the flammable solvent at a temperature that is below the flashpoint temperature? Why is our process so complex—can we simplify the process? All of these questions are related to inherently safer design. Inherently safer design should be your first approach in the prevention of fires and explosions.

Much more detail on inherently safer design is provided by Bollinger et al. (1996).

3.3. Using the Flammability Diagram to Avoid Flammable Atmospheres

Chapter 2 stated that the elimination of ignition sources is not sufficient to prevent fires and explosions—the ignition energies are too low and ignition sources too plentiful to use this as the primary prevention mechanism. A more robust design alternative is to prevent the existence of flammable mixtures as the primary control, followed by the elimination of ignition sources as the secondary control. The flammability diagram introduced in Section 2.1.1 is an important tool for determining if a flammable mixture exists, and to provide target concentrations for inerting and purging procedures presented in Section 3.4.

Section 2.1.1 showed the general shape of the flammability zone on a triangle diagram (Figure 2.7), how a flammable vapor may be created when a vessel is taken out of service (Figure 2.8), and how the flammability zone can be estimated using a few data points (Figures 2.11 to 2.13). This section will provide more details on how the flammability diagram is used in common process operations to ensure that flammable mixtures do not occur.

Consider again Figure 2.8 showing the gas composition as air is pumped into a process vessel. The objective is to avoid the flammable region altogether. The procedure is shown in Figure 3.1. The vessel is initially at point A containing pure fuel. If nitrogen is first pumped into the vessel, the gas composition follows along line AS shown in Figure 3.1. One approach would be to continue the nitrogen flow until the vessel contains pure nitrogen. However, this would require a large amount of nitrogen and would be costly. A more cost effective procedure is to inert with

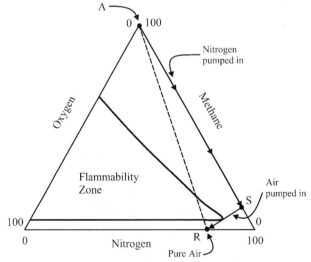

Figure 3.1. *A procedure for avoiding the flammability zone for taking a vessel out of service.*

nitrogen until point S is reached, and then including an appropriate margin of safety (NFPA 69, 1997). Then, air can be introduced, with the gas composition following along line SR on Figure 3.1. In this case the flammability zone is avoided and a safe vessel preparation procedure is ensured.

The problem now is to determine the location of point S on Figure 3.1. Clearly, the ideal approach is to have complete flammability data, such as shown in Figures 2.9 and 2.10, particularly around the nose of the flammability zone. For a number of compounds these data have been collected and the results are shown in Table 3.2 (Mashuga and Crowl, 1998). The out-of-service fuel concentration (OSFC) represents the maximum fuel concentration at point S on Figure 3.1 that just avoids the flammability zone. From a practical standpoint a fuel concentration less than the values shown in Table 3.2 would be desired to ensure that the concentrations stay well away from the flammability zone.

Figure 3.2 shows the procedure for placing a vessel into service. The vessel begins with air, shown as point A. Nitrogen is pumped into the vessel until point S is reached. Then fuel is pumped in, following line SR until point R is reached. The problem is to determine the oxygen (or nitrogen) concentration at point S. This can be done using experimental data, with the results shown in Table 3.2. The in-service oxygen concentration (ISOC) represents the maximum oxygen concentration at point S on Figure 3.2 that just avoids the flammability zone. From a practical standpoint an oxygen concentration somewhat less than the values shown in Table 3.2 would be desired to ensure that the concentrations stay well away from the flammability zone.

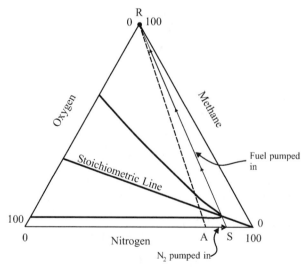

Figure 3.2. *A procedure for avoiding the flammability zone for placing a vessel into service.*

TABLE 3.2
In-Service Oxygen Concentration (ISOC) and Out of Service Fuel Concentration (OSFC) Determined from Published Experimental Data (Mashuga and Crowl, 1998)

Chemical	OSFC (Vol. % Fuel)	ISOC (Vol. % Oxygen)
Methane	14.5	13.0
Ethane	7.0	11.7
Propane	6.2	12.0
Butane	5.8	12.5
n-Pentane	4.2	12.0
n-Hexane	3.8	12.2
Natural Gas	11.0	12.8
Ethylene	6.0	10.5
Propylene	6.0	12.0
2-Methylpropene	5.5	12.5
1-Butene	4.8	11.7
3-methyl Butene	4.0	11.5
1,3-Butadiene	4.9	10.8
Acetylene	4.0	7.0
Benzene	3.7	11.8
Cyclopropane	7.0	12.0
Methyl Alcohol	15.0	10.8
Ethyl Alcohol	9.5	11.0
Dimethyl Ether	7.1	11.0
Diethyl Ether	3.8	11.0
Methyl Formate	12.5	11.0
Isobutly Formate	6.5	12.7
Methyl Acetate	8.5	11.7
Acetone	7.8	12.0
Methyl Ethyl Ketone	5.3	11.5
Carbon Disulfide	2.5	6.0
Gasoline (115/145)	3.8	12.0
JP-4	3.5	11.7
Hydrogen	5.0	5.7
Carbon monoxide	19.5	7.0

If a detailed flammability diagram is lacking, then the OSFC and ISOC can be estimated using a procedure described in Appendix A, Part B.

Direct, reliable experimental data under conditions as close as possible to process conditions is always recommended.

Care must be taken to ensure that the flammability zone for all operations is avoided by an appropriate safety margin. NFPA 69 (1997) requires a target oxygen concentration for storage vessels of at least 2% below the measured LOC, if the oxygen concentration is continually monitored. If the LOC is less than 5%, the target oxygen concentration is no more than 60% of the LOC. If the oxygen concentration is not continuously monitored, then the equipment must not operate at more than 60% of the LOC, or 40% of the LOC if the LOC is below 5%.

3.4. Inerting and Purging

The flammability diagrams presented in Section 3.3 provide target gas concentrations to avoid the flammability zone. Inerting and purging procedures are used to achieve that concentration.

Inerting and **purging** are the operations of using an inert gas to achieve a desired gas concentration in a process unit in order to render the atmosphere nonflammable. The procedures can be applied to any process unit, including pressure vessels, storage vessels, drums, pipes, pipelines, distillation columns, etc. Purging usually refers to the process of changing the gas concentration within a vessel or process, usually with either an inert gas (to decrease the oxygen concentration) or a fuel gas (to increase the fuel concentration). Inerting usually applies to maintaining an inert gas blanket in a vessel or process. The two terms are frequently used interchangeably. **Blanketing,** or **padding**, is the technique of continuously maintaining an atmosphere that is either inert or fuel rich in the vapor space of a container or vessel.

Since an inert gas is involved, care must be taken to prevent asphyxiation of workers. A single breath of an inert gas is usually adequate to cause loss of consciousness—breathing an inert gas is not the same as holding one's breath. Precautions must be taken to ensure that the inert gas does not escape to worker areas and that workers follow proper confined space entry procedures prior to entering any enclosed space containing an inert atmosphere.

Nitrogen is by far the most common gas used for inerting and purging procedures. However, carbon dioxide, argon, combustion product gases (such as diesel exhaust) and even water vapor are sometimes used. Water vapor is used only on a limited basis because it has the disadvantage that the vapor will condense if the temperature drops, causing the loss of the inert blanket. The shape of the flammability zone on the triangular diagram, and thus the flammability limits, the

MOC, etc. change with the type of inert gas. The availability of specific data for the inert gas selected must be considered in the selection.

Many monomers (such as styrene and acrylic acid) require inhibitors to prevent polymerization of the polymer liquid in storage. Some of these inhibitors require the presence of a small amount of oxygen to be effective. The inhibitor effectiveness would be lost under an entirely oxygen free inert blanket. This is an important consideration in the design of inerting systems for such monomers.

Several different inerting and purging procedures are available. The selection of a specific procedure will depend on the design of the process equipment, the utilities available, the type and cost of the inert, the vacuum or pressure rating of the process, the geometric configuration of the process and the time available to perform the procedure.

3.4.1. Vacuum Purging

Vacuum purging is used if a vacuum system is available and the process equipment can withstand the effects of vacuum. This procedure involves evacuating the process followed by filling with an inert gas, with the procedure being repeated until the target gas concentration is achieved.

Suppose that a process vessel must be inerted using a vacuum purging method. The volume of the vessel is known. The oxygen concentration must be reduced from an initial concentration, y_0, to a final target oxygen concentration, y_j. The cycles used to accomplish this procedure are shown in Figure 3.3. The vessel is initially at absolute pressure P_H and is vacuum purged using a vacuum at abso-

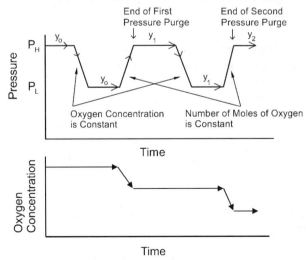

Figure 3.3. Pressure and oxygen concentration histories for vacuum purging with pure inert gas.

lute pressure P_{L}. The objective of the calculation is to determine the number of cycles required to achieve the desired oxygen concentration.

During evacuation of the vessel the composition of the gas within the vessel remains constant. Thus, the oxygen mole fraction remains constant, but the number of moles of oxygen is decreased. The vessel is repressurized back to P_{H} using pure inert. During this operation, the number of moles of oxygen remains constant and the mole fraction decreases.

The vessel initially has an oxygen mole fraction of y_0. At the completion of the first evacuation, the number of moles of oxygen present is

$$n_{\text{oxy}} = y_0 \left(\frac{P_{\text{L}} V}{R_{\text{g}} T} \right) \tag{3-1}$$

where n_{oxy} is the number of moles of oxygen (moles)
　　　y_0　is the initial mole fraction concentration of oxygen (unitless)
　　　P_{L}　is the low pressure vacuum, in absolute pressure units (pressure)
　　　V　is the volume of the process or vessel (volume)
　　　R_{g}　is the ideal gas constant (pressure-volume/mole-deg)
　　　T　is the absolute temperature (deg)

This is also the total number of moles of oxygen present at the end of the first repressurization.

At the completion of the first repressurization, the total number of moles present is

$$n_{\text{tot}} = \frac{P_{\text{H}} V}{R_{\text{g}} T} \tag{3-2}$$

where n_{tot} is the total number of moles of gas (moles) and P_H is the high, or ambient pressure, in absolute pressure units (pressure).

The resulting mole fraction of oxygen, y_1, at the completion of the first vacuum cycle is thus,

$$y_1 = \frac{n_{\text{oxy}}}{n_{\text{tot}}} \frac{y_0 (P_{\text{L}} V / R_{\text{g}} T)}{P_{\text{H}} V / R_{\text{g}} T} = y_0 \left(\frac{P_{\text{L}}}{P_{\text{H}}} \right) \tag{3-3}$$

This is also the mole fraction of oxygen at the completion of the second evacuation.

At the completion of the second repressurization, the oxygen mole fraction, y_2, is given by

$$y_2 = y_1 \left(\frac{P_{\text{L}}}{P_{\text{H}}} \right) = y_0 \left(\frac{P_{\text{L}}}{P_{\text{H}}} \right)^2 \tag{3-4}$$

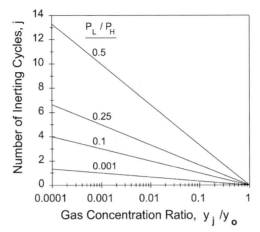

Figure 3.4. Plot of Equation (3-5) showing the number of cycles required for inerting.

Continuing in this fashion, the mole fraction of oxygen at the end of the *j*th vacuum cycle is given by,

$$y_j = y_0 \left(\frac{P_L}{P_H} \right)^j \tag{3-5}$$

Figure 3.4 is a plot of Equation (3-5). Equation (3-5) assumes that the gas is well-mixed between each cycle.

Fractional cycles determined using Equation (3-5) would normally be rounded up. Thus, if 1.2 cycles is calculated, at least two cycles would be used. The actual number of cycles used would normally be greater than the number computed, to provide a reasonable safety margin.

The total moles of inert used, Δn_{inert} during the inerting procedure is given by the ideal gas law,

$$\Delta n_{inert} = j(P_H - P_L) \frac{V}{R_g T} \tag{3-6}$$

3.4.2. Pressure Purging

Pressure purging can be used if the process equipment is designed to withstand pressure. This procedure involves pressurizing the process with inert gas followed by venting the process back to ambient pressure. The procedure is repeated until the target gas concentration is achieved.

This procedure is shown in Figure 3.5. In this case, the vessel is initially at P_L and is pressurized using a source of pure nitrogen at P_H. The objective is to determine the number of cycles required to achieve a desired oxygen concentration.

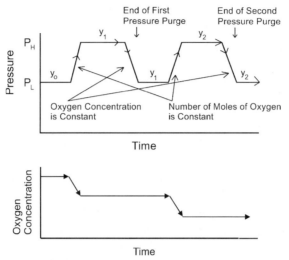

Figure 3.5. *Pressure and oxygen concentration histories for pressure purging with pure inert gas.*

Since the vessel is pressurized with pure nitrogen, the number of moles of oxygen remains constant during pressurization while the mole fraction decreases. During depressurization, the composition of the gas within the vessel remains constant, while the total number of moles is reduced. Thus, the oxygen mole fraction remains unchanged.

Equation (3-3) provides the oxygen mole fraction at the end of the first pressurization. This is also the oxygen mole fraction at the end of the first de-pressurization (the end of the first pressure cycle). It is readily shown that Equation (3-5) provides the general result.

3.4.3. Combined Pressure–Vacuum Purging

In some facilities, a pressurized inert source and vacuum are available and can be used simultaneously to purge a vessel. **Combined pressure–vacuum purging** is performed by first evacuating the process, followed by pressurization with the inert gas. The procedure is repeated until the target concentration is achieved.

This procedure is shown in Figure 3.6. The first evacuation cycle occurs between the initial pressure, P_0, and the final pressure P_L. During the evacuation the moles of oxygen are reduced and the oxygen concentration remains constant. The oxygen mole fraction at this point is the same as the initial mole fraction. The number of moles of oxygen is given by Equation (3-1) and is dependent only on the final pressure, P_L, and not on P_0. Furthermore, the remaining cycles are identical to the vacuum purge operation and Equation (3-5) is applicable. This case is

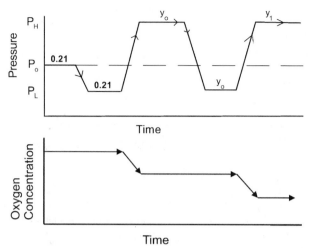

Figure 3.6. *Combined pressure–vacuum purge—evacuate first.*

nearly identical to the vacuum purging case considered earlier, with the exception of the first cycle.

The procedure can also be performed by pressurizing first, as shown in Figure 3.7. For this case, during the first pressurization, between P_0 and P_H, the oxygen concentration decreases but the number of moles of oxygen remains constant. If the initial oxygen mole fraction is 0.21, the oxygen mole fraction at the end of this initial pressurization is given by

$$y_0 = 0.21\left(\frac{P_0}{P_H}\right) \tag{3-7}$$

After the first pressurization the remaining cycles are identical to pressure purging alone and Equation (3-5) applies. However, the number of cycles, j, determined from Equation (3-5) is the number of cycles after the initial pressurization. Thus, the total number of cycles is $j + 1$.

Pressurizing first on a combined pressure–vacuum purge requires more inert gas and more inerting cycles to achieve the desired result. By pressurizing on the first cycle, the moles of oxygen in the vessel remains constant. The reduction in inert gas usage and cycles by evacuating first is achieved by the reduction in the number of moles of oxygen in the vessel during the first initial evacuation.

The equations developed for vacuum and pressure purging apply for the case of pure inert only. Many inert gas separation processes available today do not provide pure inert—they typically provide inert in the 98%+ range. The equations provided for pure nitrogen can be modified to account for this. Assume that the

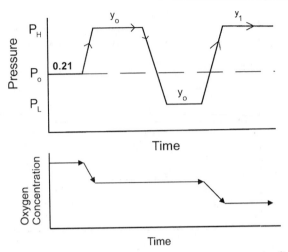

Figure 3.7. Combined pressure–vacuum purge—pressurize first.

inert contains oxygen with a constant mole fraction of y_{oxy}. For a pressure purging procedure, the total moles of oxygen present at the end of the first pressurization is given by the moles initially present plus the moles included with the inert. This amount is

$$n_{oxy} = y_0 \left(\frac{P_L V}{R_g T} \right) + y_{oxy} (P_H - P_L) \frac{V}{R_g T} \tag{3-8}$$

The total moles in the vessel at the end of the first pressurization is given by Equation (3-2). Thus, the mole fraction of oxygen at the end of this cycle is

$$y_1 = \frac{n_{oxy}}{n_{tot}} = y_0 \left(\frac{P_L}{P_H} \right) + y_{oxy} \left(1 - \frac{P_L}{P_H} \right) \tag{3-9}$$

This result is generalized into the following recursive equation for the oxygen concentration at the end of the *j*th pressure cycle,

$$y_j = y_{j-1} \left(\frac{P_L}{P_H} \right) + y_{oxy} \left(1 - \frac{P_L}{P_H} \right) \tag{3-10}$$

Equation (3-10) is used in place of Equation (3-5) for both pressure and vacuum purging.

3.4.4. Sweep Purging

Sweep purging can be used if the process equipment is unable to withstand the effects of pressure or vacuum, such as a storage tank. This involves using a continuous flow of inert through the process until the desired target concentration is achieved.

Sweep purging calculations are performed assuming perfect mixing within the vessel, constant temperature and constant pressure. Thus, the volumetric flow of gas out of the vessel is equal to the flow in, and the outlet gas concentration is equal to the gas concentration within the vessel. A total mass balance around the vessel results in

$$V\frac{dC}{dt} = C_0 Q_v - CQ_v \tag{3-11}$$

where V is the vessel volume (volume)

C is the concentration of oxygen within the vessel (mass/volume)

C_0 is the oxygen concentration in the inlet inert gas (mass/volume)

Q_v is the volumetric flow rate of inert (volume/time).

Equation (3-11) can be rearranged and integrated from initial oxygen concentration, C_1, to a target concentration of C_2. A nonideal mixing factor, K, can also be introduced to account for the fact that the mixing will not be perfect.

$$Q_v t\left(\frac{V}{K}\right)\ln\left(\frac{C_1 - C_0}{C_2 - C_0}\right) \tag{3-12}$$

The quantity $Q_v t$ represents the total volumetric flow of inert required to meet the target concentration. Once the volumetric flow of inert, Q_v is specified, then the time required to obtain the target concentration is determined. A value for K no larger than 0.25 is recommended by NFPA 69 (1997).

3.4.5. Siphon Purging

Siphon purging is an inerting procedure in which the process is filled with a compatible fluid, such as water. The liquid is subsequently drained out, with the gas being replaced by inert gas as the liquid is drained. The total quantity of inert used is equal to the volume of liquid displaced. The liquid used must be readily available and must be compatible with the materials of construction and the material ultimately stored in the vessel. Water is frequently used.

Many storage vessels cannot be filled completely with liquid. In this case the vessel is filled as far as possible with the liquid, then a sweep purging method is used to inert the remaining volume. In many cases, the gas in this head space represents such a small part of the total volume that it is negligible.

3.4.6. Advantages and Disadvantages of the Various Inerting Procedures

Pressure purging is generally faster since the pressure differentials are greater, however, it uses more inert gas than vacuum purging. Vacuum purging uses less inert gas since the oxygen concentration is reduced primarily by vacuum. For combined vacuum–pressure purging, less nitrogen is used if the initial cycle is a vacuum cycle. If a process geometry is complex, with many connections and closed pipe sections, a sweep purging method may be impractical, and either a vacuum, pressure or combined method will be necessary.

3.4.7. Inert Gas Blanketing of Storage Vessels

Two methods are used to prevent flammable vapors in storage vessels: continuous sweep and static blanketing methods. In the **continuous sweep** method, inert gas flows continually through the vessel. In the **static blanketing** method, inert gas is provided at a fixed pressure, typically 0.5 inches of water gauge.

In the past, storage vessels were protected mostly by the continuous sweep method. Today, due to the cost of inert gas and due to fugitive emissions reduction requirements, static blanketing methods are more popular. For both methods the inert supply system must be capable of handling normal pressure and temperature changes in the vessel, external fire, and changes due to pumping process liquid into and out of the vessel. Details on gas blanketing sizing are provided in API 2000 (1998).

The static blanketing method uses a conservation vent and a gas regulator. The conservation vent maintains a low backpressure and opens and discharges vapor when the pressure exceeds a set value. The gas regulator system provides make-up inert gas when the pressure drops. The conservation vent also frequently provides vacuum relief in the event of static blanketing system failure.

The gas regulator system for static blanketing systems must be carefully designed due to the low pressures considered (typically 0.5 inches of water gauge, or about 125 Pa). Most spring-loaded regulators (including typical low pressure devices) are designed to regulate the pressure in a flowing stream and are not capable of providing the accuracy or response for static pressure regulation. Typical regulator systems for inert gas blanketing use a pilot sensing regulator to achieve the proper function (Richard, 1986).

3.4.8. Inert Purging and Blanketing during Drumming Operations

Drums are used frequently to charge materials into a process and to collect and store products or waste materials. Air can enter the drum during these operations, forming a flammable atmosphere, and presenting a fire and/or explosion hazard. Also, air may initially be in the ullage of the drum. Procedures are available to inert the drum for both charging and drumming operations.

Figure 3.8 shows a charging operation where the contents of the drum is being charged into a process. Without inerting, air will be drawn into the drum as the liquid is removed and the liquid level drops. To prevent this, the drum is fitted with a "T" fitting, as shown. One end of the "T" is connected to a low pressure inert gas

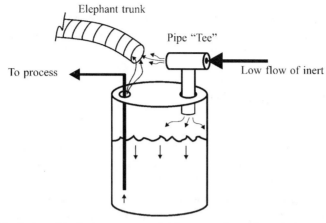

Figure 3.8. *An inert blanketing system useful for protecting a drum during process charging.*

source with a small, but continuous flow. The other end of the "T" is open to prevent the accidental overpressuring of the drum by the inert gas source. If the operation is being performed in an enclosed work area, then an elephant trunk, canopy hood, or other ventilation source must be provided to remove the inert gas to prevent worker asphyxiation. Now, as the liquid level moves downward, the vapor space is filled with inert gas, preventing the formation of a flammable vapor.

Figure 3.9 shows an inerting procedure for an empty drum prior to filling from a process. In this case a wand long enough to reach the bottom of the drum is placed into the drum and a flow of low-pressure inert is initiated. If the operation is being performed in an enclosed work area, then an elephant trunk, canopy hood, or other ventilation source must be provided to remove the inert gas to prevent worker

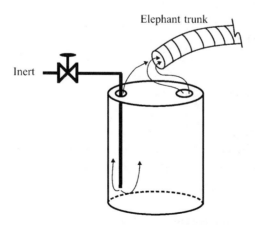

Figure 3.9. *Inerting a drum prior to filling with a flammable liquid.*

asphyxiation. After a suitable length of time, defined as the time required to flush the drum with three to four times the drum volume with inert gas, the wand is removed and the vessel is ready for filling.

3.5. Example Application

EXAMPLE 3.1
A 100-gallon reactor vessel must be inerted with nitrogen prior to charging with methyl ethyl ketone (MEK). Pure nitrogen is available at 20 psig and a vacuum system is available at 60 mm Hg abs. Determine:

   a. The target oxygen concentration for the inerting procedure.
   b. The number of cycles required for a combined pressure—vacuum inerting procedure and the total quantity of nitrogen used.
   c. The time required to sweep purge the vessel assuming the nitrogen flow from the inert system is 20 CFM.

Assume the vessel is initially filled with air at 1 atm and 80°F.

Solution a: From Table 3-2, the in-service oxygen concentration for MEK is 11.5% oxygen. The oxygen concentration must be reduced below this value. For a safety margin use 60% of this value, giving a target oxygen concentration of 6.9%.

If we use Equation (A-16) to estimate the in-service oxygen concentration, we need to determine z, the stoichiometric coefficient for the combustion reaction and the LOC. For MEK, the combustion reaction is:

$$C_4H_8O + 5.5\ O_2 \rightarrow 4\ CO_2 + 4\ H_2O$$

Thus, $z = 11/2 = 5.5$. From Table 2.3 the LOC for MEK is 11% oxygen. From Equation (A-16)

$$ISOC = \frac{z * LOC\%}{z - \dfrac{LOC\%}{100}} = \frac{5.5 * 11}{5.5 - \dfrac{11}{100}} = 10.2\%$$

This value is somewhat lower than the experimental value of 11.5% from Table 3-2.

Solution b: For the combined pressure–vacuum purging method we will evacuate first to reduce the oxygen concentration as much as possible. Assume the evacuation is terminated at 100 mm Hg to reduce the evacuation time. Thus,

$$P_H = 20\ \text{psig} + 14.7\ \text{psi} = 34.7\ \text{psia}$$

$$P_L = (100\ \text{mm Hg})\left(\frac{14.7\ \text{psia}}{760\ \text{mm Hg}}\right) = 0.132\ \text{psia}$$

Substituting into Equation (3-5),

$$y_j = y_0 \left(\frac{P_L}{P_H} \right)^j$$

$$7.5\% = (21\%) \left(\frac{0.132 \text{ psia}}{34.7 \text{ psia}} \right)^j$$

$$\ln\left(\frac{7.5}{21} \right) = j \ln\left(\frac{0.132}{34.7} \right)$$

$$j = \frac{-1.030}{-5.572} = 0.185$$

Thus, one cycle is more than adequate to reach the target oxygen concentration.

The total nitrogen used is provided by Equation (3-6). Assuming a single inerting cycle,

$$\Delta n_{\text{inert}} = j(P_H - P_L)\frac{V}{R_g T}$$

$$= (1 \text{ cycle})(34.7 \text{ psia} - 0.132 \text{ psia}) \frac{(100 \text{ gal})(0.1337 \text{ ft}^3/\text{gal})}{\left(10.731 \dfrac{\text{psia ft}^3}{\text{lb - mole } {}^\circ\text{R}} \right)(80 + 140 \text{ R})}$$

$$\Delta n_{\text{inert}} = 0.0797 \text{ lb-moles}$$

The total mass of nitrogen is $(0.0797 \text{ lb-moles})(28 \text{ lb/lb-mole}) = 2.2 \text{ lb}$ of nitrogen.

Solution c: Equation (3-12) applies for a sweep purge. Since the nitrogen is pure, $C_0 = 0$. Assume a nonideal mixing factor of 0.25. Substituting into Equation (3-12)

$$Q_v t\left(\frac{V}{K} \right) \ln\left(\frac{C_1}{C_2} \right)$$

$$Q_v t = \left(\frac{100 \text{ gal}}{0.25} \right)\left(\frac{0.1337 \text{ ft}^3}{\text{gal}} \right) \ln\left(\frac{21}{7.5} \right) = 55.1 \text{ ft}^3$$

The total inerting time would be

$$t = \frac{55.1 \text{ ft}^3}{Q_v} = \frac{55.1 \text{ ft}^3}{20 \text{ ft}^3/\text{min}} = 2.75 \text{ min}$$

This represents 3.91 lb of nitrogen. As expected, the sweep purge requires more nitrogen than the combined pressure–vacuum purge.

3.6. Explosion Venting

Explosion venting can be used to lower the maximum pressure developed by a deflagration, for example, where the flame speed is lower than the speed of sound, but is not effective against detonations. The vent area required to prevent the explosion pressure from exceeding a desired value is dependent on several variables, including:

- The properties of the materials present, which are characterized by their deflagration index, K_G for gases and K_{St} for dusts. See Equations (2-1) and (2-22), respectively.
- The vent opening pressure.
- The length to diameter ratio (L/D) and volume of the enclosure being vented.
- The weight of the vent panel (which should normally be below 2.5 lb_m/ft^2).

Other special circumstances must also be considered:

- If the enclosure is vented through a duct the vent area requirements will increase.
- If the vessels are interconnected, pressure piling (described in Section 2.7.4) can significantly increase the vent area.

Additional details on this subject are provided by NFPA 68 (2002).

3.7. Grounding and Bonding

Grounding and bonding procedures are used to prevent the accumulation of static charge in order to prevent the ignition of a flammable atmosphere due to static discharge. These procedures are also important to protect against electrical faults and lightning. **Grounding** is the procedure of connecting process equipment and materials to an electrical ground in order to remove any accumulated charge. **Bonding** is the procedure of connecting two process units together electrically in order to maintain the same electrical potential between the two units. Grounding and bonding is effective only for conductive components and materials.

Static electricity may accumulate in both process equipment and materials within the process. Grounding and bonding methods must be employed to dissipate the static electricity from both. Process materials may include liquids, solids and gases. Grounding and bonding of process materials can be difficult, particularly for solids and gases, and is normally achieved through the process equipment or through grounding rods placed into the material.

Many facilities have a dedicated grounding system for dissipating static charge. This ground is separate from the electrical ground in order to prevent energizing of process equipment in the event of an electrical ground fault, or loss of ground due to changes in the wiring system. In order to obtain a low enough resistance to ground, the ground system must cover a large area and should consist of either a buried wire mat or a system composed of many interconnected grounding rods. A single ground rod, such as used for a lightning protection system or a residential ground, is usually not adequate.

If lightning ground rods are used, NFPA 780 (1997) suggests that the rods should be at least ½-inch in diameter and 8-feet long. The rod should be of copper-clad steel, solid copper, hot dipped galvanized steel or stainless steel and free of paint or other nonconductive coatings. The 8-foot length might present a problem for temporary ground systems during emergency field transfers of liquids or solids from trucks or railroad cars in the field.

A ground resistance test meter is an electronic device used to determine the resistance from the ground rod to the physical ground. The device injects a current between a temporary, remote test probe and the ground system. A second temporary test probe is used to measure the resultant potential created by the test current. From the voltage and current values the resistance can be calculated. Permanent ground systems must be checked on a regular basis and maintained to ensure their integrity. A nominal resistance between the ground system and the physical ground of no more than 1000 ohms is desirable to dissipate static charge. This resistance could be as high as several megohms and still dissipate static charge, but sparks to adjacent grounded equipment may occur.

Resistances between the grounding system and process equipment can be as high as a megohm to dissipate static charge. However, from a practical standpoint, a resistive value of 6 or 7 ohms is used to ensure a good mechanical connection for grounding and bonding connections in a process plant.

Figure 3.10 shows a typical grounding and bonding arrangement. In this example a metal container is bonded to a grounded, unlined, metal drum prior to dispensing liquid from the drum to the container. The container is also grounded, as shown. Two types of clamps are used. One type is for longer-term installations and is composed of a threaded, sharp-pointed screw on a clamp. The other type is for temporary operations, for instance, to bond or ground a drum to a process. This is a spring loaded clamp that can be readily attached to the drum by hand. In both types of clamps, a metal point penetrates any paint that might be present on the surface to ensure contact with metal. Common electrical alligator clips must not be used for grounding and bonding since the alligator clip spring or points might not be adequate to penetrate any paint layers. In Figure 3.10 the spring-loaded clamp is used to bond the container to the drum, while the bolt clamps are used to connect the bond and the ground to the drum. NFPA 77 (2000) provides more detail on grounding and bonding.

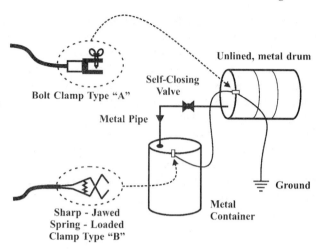

Figure 3.10. *Bonding and grounding procedures for tanks and vessels (Eichel, 1967).*

Figure 3.11 shows how an unlined, metal drum or a metal tank truck or car is grounded to the process. In the upper part of Figure 3.11 the drum is bonded to the process, the process is grounded, and the transfer line is bonded at both ends to the conductive hose. In the lower part of Figure 3.11, the tank truck or car is bonded to the process and grounded. The process itself must also be grounded. Finally, the conductive hose is bonded to the piping at both ends.

If the drum in Figure 3.10 were lined with a nonconductive material, such as plastic, then a different approach must be used since the static which may accumulate in the liquid has no pathway to ground. For this case the drum must be placed in the vertical position and a metal rod must be placed into the liquid through an open bung hole. The metal rod is then bonded to the process. Care must be taken to ensure that the metal rod does not penetrate the drum lining at the bottom of the drum. For the drum shown in the upper part of Figure 3.11, the metal dip pipe would provide the necessary pathway to dissipate static electricity in the event the drum were lined.

Many materials are now transported in plastic totes, with volumes up to several hundred gallons. These totes are constructed from a plastic tank encased in a metal support structure or cage. A metal rod must be placed into the liquid through the filling hole at the top of the tote, and grounded or bonded to remove static charge. The metal supporting cage or structure must also be grounded or bonded.

For tank trucks or railway tankers special care must be taken in attaching the bonding wire to the truck or tanker. Many trucks and railway tankers have metal components, such as the wheels or supporting carriage, that are insulated from the tank and do not provide an adequate pathway for dissipation of static from the tank

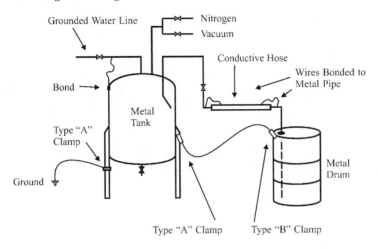

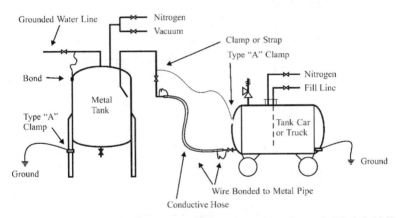

Figure 3.11. Bonding and grounding procedures for tanks and vessels (Eichel, 1967).

contents. In general, bolted on metal pieces, handrails or valve handles may not provide adequate bonding or grounding points. Metal pieces welded directly to the tank provide better grounding and bonding points. Also, heavy paint or corrosion might present a challenge—the grounding area should be sanded down to bare metal to provide good contact. For double bottom (tandem) tankers (two tank trailers hooked together and towed by one tractor) both tanks must be bonded together.

If a temporary ground system must be made, for instance, for a field transfer of liquid between tank trucks, then care must be taken to construct the ground properly. A single rod driven into the ground may not provide adequate grounding, since it typically will have a ground resistance of several thousand ohms. A better approach is to dig a small depression several inches deep and about 6 feet in diameter. Drive several grounding rods into this area, and then fill the depression with water and add salt to increase the conductivity of the water. All of the rods are connected together to provide the ground. This type of system typically has a ground resistance of less than 100 ohms—even in porous sand conditions.

All conductive process equipment must be bonded and grounded. This includes any isolated pieces of metal that are large enough to have adequate capacitance to accumulate enough charge for a static discharge. Some companies have corporate standards specifying a maximum metal surface area above which bonding is required. A typical value used is 10 cm^2. Of particular problem are flanges with nonconductive gaskets, hoses, bearings, and other devices within the process. Figure 3.12 shows a few methods to handle some of these problems.

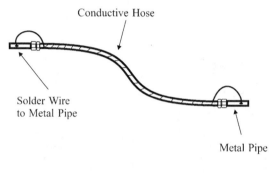

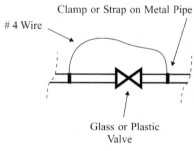

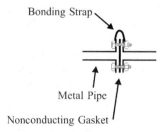

Figure 3.12. Bonding procedures for valves, pipes, and flanges (Eichel, 1967).

Process equipment constructed of glass, teflon, plastic, or other types of nonconductive material or lining presents a challenge with respect to grounding and bonding. Figure 3.13 shows a tantalum insert into a glass reactor to provide a ground for the conductive liquid inside the reactor.

Piping, containers, storage vessels, pumps and other process equipment constructed of plastic can be made conductive by using conductive additives, such as carbon. However, the use of these materials presents a management problem since it is difficult to discern the difference between the conductive and nonconductive plastics. As a result, many facilities that handle flammable solvents forbid the use of any type of plastic (or other nonconductive) equipment in these areas.

Free fall of liquids or powders into a vessel can result in considerable charge accumulation. Electrostatic charging results from the separation of liquid from the pipe and from droplet-droplet separation. For liquids, a dip tube may be used to dissipate the static charge, as shown in Figure 3.14. The dip tube also provides a

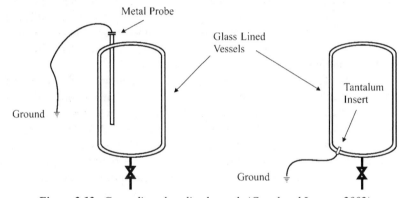

Figure 3.13. *Grounding glass-lined vessels (Crowl and Louvar, 2002).*

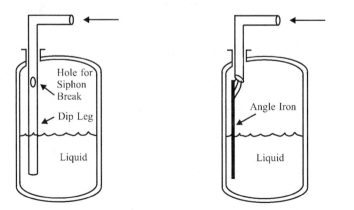

Figure 3.14. *Dip legs to prevent free fall and accumulation of static charge (Crowl and Louvar, 2002).*

means to dissipate charge from the liquid in the vessel. A small hole must be provided in the dip tube to prevent siphoning of the liquid back out of the vessel. Another approach is to provide a V-shaped metal angle iron for the liquid to fall down, as shown in Figure 3.14.

Large flexible intermediate bulk containers (FBICs) that contain up to 1 ton of powder present a problem with respect to charge dissipation. These are discussed in more detail in Section 3.12.

3.8. Ventilation

Ventilation is important not only for removing toxic materials from the workplace but also for reducing the concentrations of flammable or combustible materials below flammable levels. For flammable or combustible materials, the ventilation depends on the type of material handled. Figure 3.15 shows the major classifications for flammable and combustible liquids, as defined by NFPA 30 (NFPA 30, 2000). The classification scheme depends mostly on the flashpoint of the liquid. However, for IA and IB liquids, the classification is further subdivided by the boiling point of the liquid.

Two types of ventilation methods are used: dilution and local ventilation. **Dilution ventilation** involves adding fresh air into a work area to reduce the concentrations of flammable or toxic materials. The problems with this approach are (1) the workers are continuously exposed to the material; and (2) the heating and cooling energy requirements can be large. For flammable materials, the dilution air required to reduce the concentrations below the lower flammability limit may be small or large, depending on the material. **Local ventilation** involves the use of

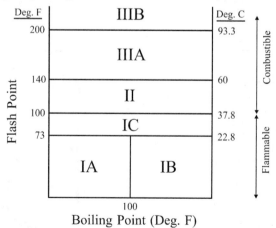

Figure 3.15. Classification for flammable and combustible liquid as defined by NFPA 30 (2000).

hoods, elephant trunks, or canopies to capture the material at the point of evolution. The advantage to this approach is that worker exposures are minimal. If a hood is used it may restrict equipment access and can only be used for a small process. Design methods and calculations for ventilation systems are provided elsewhere (ACGIH, 1998; Crowl and Louvar, 2002).

For inside storage rooms with either a gravity or mechanical ventilation system, OSHA recommends changing the air within the room a minimum of six times per hour. For instance, if the room volume were 1000 ft^3, then the ventilation system must provide a minimum of 6×1000 ft^3 = 6000 ft^3/hour = 100 ft^3/min. If a mechanical exhaust system is used, then it must be controlled by a switch outside the room, with the ventilation system and lights being controlled by the same switch. An indicator light must be installed adjacent to the switch if Class I flammables are dispensed within the room (OSHA, 1996). NFPA 30 (2000) recommends a minimum ventilation rate for inside storage rooms of 1 ft^3/min of air per ft^2 of floor area, but not less than 150 cfm. The electrical system within the room must meet the requirements for electrical systems in classified locations (see Section 3.13).

For processing areas inside a building, the ventilation system must provide a minimum of 1 ft^3/min per ft^2 of floor area. Thus, for a facility with 5000 ft^2 of total floor space, the minimum ventilation rate would be 5000 ft^3/min. The electrical system within the processing area must meet the requirements for electrical systems in rated locations (see Section 3.13). The ventilation system must discharge the exhaust to a safe location outside the building without short-circuiting with the intake air. The ventilation system must also include all floor or pit areas where flammable vapor may collect (OSHA, 1996).

3.9. Sprinkler and Deluge Systems

A **sprinkler system** is a network of piping and discharge nozzles throughout a structure or process area through which water is discharged during a fire. The water quenches the fire and also cools and protects the equipment and structures from the fire. The complete details on sprinkler system design is beyond the scope of this book and are provided elsewhere (NFPA 13, 1999; NFPA 15, 2001).

The discharge heads or nozzles are generally located high in the building or process area and are activated by a number of mechanisms. Some systems are continuously pressurized with water with the heads activated individually by the melting of a fusible link or other mechanism at the sprinkler head. This is called a closed-head system. Activated heads cannot be turned off once activated unless the water flow is stopped. Closed-head systems are used in storage areas, laboratories, and small pilot plant areas. Another type of system contains open sprinkler heads, with water being present in the system only after detection of the fire and activation of the water flow from a centralized point. This is called an open-head system. All of the sprinkler heads in an open-head system are activated once the water flow is started. This provides additional protection to adjacent areas and is suitable for plant process areas and larger pilot plants. The fire detection system for an open-head sprinkler system can be based on a number of approaches, including heat, smoke, or flammable vapor detectors. The system can also be activated manually.

Some concern is always raised about water damage, particularly in computer rooms, control rooms or buildings. Generally, the water damage from a sprinkler system is always less than the damage caused by a fire without a sprinkler system. For larger plant areas, containment might be necessary to hold the large quantities of water produced to prevent runoff and potential environmental damage to adjacent areas.

Deluge systems are typically open-head systems designed to discharge large quantities of water on to the surface of a process unit, such as a storage vessel, reactor, or even a tank truck. The main purpose of the deluge system is to keep the surface of the designated unit cool and protect the unit from fire exposure. However, the deluge system also flushes away potentially hazardous materials, and helps to knock down any gas clouds. Deluge systems are frequently manually activated to protect process equipment from the thermal effects of fires in adjacent units. Deluge systems may also be activated using flammable vapor detectors, which typically activate at 25% of the LFL. Another means of activation are air pilot systems with a fusible link that melts when exposed to fire, losing air pressure and activating the deluge. Finally, smoke and heat detectors are also used.

Fire monitor nozzles are usually installed at a fixed location and consist of a manually activated and adjustable water nozzle which produces a continuous stream reaching a maximum distance of about 150 feet. The resulting stream can be adjusted by flow, angle and direction, providing the capability to target a particular location in the process.

Sprinkler and deluge systems and fire monitor nozzles must be properly maintained and the water supply must be adequate and reliable. Many facilities use a dedicated water storage tank and diesel-powered pumps to provide water for sprinkler protection. The diesel engine is started when a fire is detected, causing a delay of a few minutes in activation of the sprinkler or deluge system until the diesel engine and pump comes up to speed and the deluge piping is filled with water. Fire pumps, whether diesel or electric motor driven, must be run each week to verify proper starting and operation. NFPA 25 (2002) addresses the proper testing procedures for fire pumps.

NFPA 13 (1999) describes several different hazard occupancies, with the classification depending on the quantity and combustibility of the materials used. The sprinkler system design depends on the hazard occupancy. Most chemical plants using flammable solvents are classified as an Ordinary Hazard (Group 3) occupancy. This means that the quantity and/or combustibility of the contents is high and fires with a high rate of heat release are expected.

Table 3-3 provides a summary of the design requirements for sprinkler systems in Ordinary Hazard (Group 3) occupancy classifications. In 2002, NFPA changed Ordinary Hazard (Group 3) to Extra Hazard (Group 2). The sprinkler system design—including head placement, pipe and pump sizing—is done using detailed hydraulic calculations, including pipe frictional losses, elevation changes, and pump efficiencies. Where flammable or combustible liquids are stored in significant quantities, NFPA 30 (2000) addresses storage arrangement and sprinkler layout.

TABLE 3.3
Sprinkler Protection for Chemical Plants Based on Ordinary Hazard (Group 3)[a]
(Crowl and Louvar, 2002)

SPRINKLER SYSTEM TYPES:

• Antifreeze sprinkler system: a wet pipe system that contains an antifreeze solution and that is connected to a water supply.

• Deluge sprinkler system: open sprinklers and an empty line that is connected to a water supply line through a valve that is opened upon detection of heat or a flammable material.

• Dry pipe sprinkler system: A system filled with nitrogen or air under pressure. When the sprinkler is opened by heat, the system is depressurized, allowing water to flow into the system and out the open sprinklers.

• Wet pipe sprinkler system: a system containing water that discharges through the opened sprinklers via heat.

DESIGN DENSITIES (see NFPA 13, 1999; NFPA 15, 2001)

• *Source of fire:* not less that 0.50 gpm/ft^2 of floor area.

• *Pumps and related equipment:* 0.50 gpm/ft^2 of projected area.

• *Vessels:* 0.25 gpm/ft^2 of exposed surface, including top and bottom. Vertical distance of nozzle should not exceed 12 ft.

• *Horizontal structural steel:* 0.10 gpm/ft^2 of surface area. This may not be necessary if the steel is insulated or designed to withstand the worse-case scenario.

• *Vertical structural steel:* 0.25 gpm/ft^2 of surface area. This may not be necessary if the steel is insulated or designed to withstand the worst-case scenario.

• *Metal pipe, tubing, and conduit:* Not less than 0.15 gpm/ft^2 of surface area and directed toward the undersides.

• *Cable trays:* Not less than 0.3 gpm/ft^2 of projected plane area (horizontal and vertical).

• *Combined systems:* The NFPA standards specify acceptable methods of combining the above requirements.

• *Nominal discharge rates for 0.5-in. orifice spray nozzles:*

gpm:	18	25	34	50	58
psi:	10	20	35	75	100

---

[a]In 2002 the NFPA renamed Ordinary Hazard (Group 3) as Extra Hazard (Group 2).

3.10. Charging and Drumming Flammable Liquids

Toxic and flammable materials are charged and drummed from process equipment on a regular basis and in large quantities. The hazardous properties of the material must be fully understood in order to develop a safe procedure for handling this material.

The proper charging and drumming procedure for flammable liquids is based on the following fundamental concepts:

1. The primary defense against fires/explosions is to prevent the existence of flammable atmospheres.
2. The secondary defense against fires/explosions is to eliminate ignition sources.
3. Worker exposures to the chemical must be minimized.

These methods are best illustrated by an example.

3.11. Example Application

EXAMPLE 3.2

Develop a standard operating procedure (SOP) for (a) vacuum charging methyl ethyl ketone (MEK) into the reactor vessel shown in Figure 3.16 from 55-gallon drums, and (b) transferring the processed material back into 55-gallon drums. Procedures involving inerting and purging, ventilation, and personal protection must be followed.

The properties of the material are provided by the MSDS sheet and are summarized below.

The vessel, as found, is empty and washed. It contains air.

The 55-gallon drum containing the liquid material is initially closed and has two bung holes. The Quality Control lab has confirmed that the methyl ethyl ketone in the drum is as stated on the MSDS.

MSDS Information on MEK
   Flashpoint: 20°F (-6.7°C)
   LFL: 1.8%
   UFL: 10.0
   Boiling point: 176°F (80°C)
   TLV-TWA: 200 ppm

MEK is a stable material in closed containers at room temperature under normal storage and handling conditions.

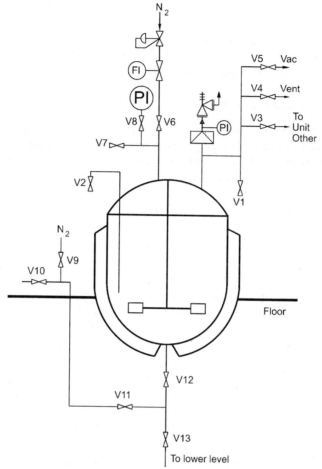

Figure 3.16. Reactor system for Example 3.2.

Process Equipment Information

Vessel:
- 100 gallon, jacketed, glass-lined vessel, 30" ID × 41.25" deep, sight glass equipped.
- Pressure rated at 25 psig at 650°F
- 2-inch safety valve set at 25 psig, maintained within year
- 2-inch 316 stainless steel rupture disc set at 22 psig, maintained within the last year.
- Hydrostatic tested at 25 psig within the year.
- Process is grounded to building ground system, checked within year.

Drive:
- Variable speed belt drive, 10 to 1 ratio to give 60 to 250 rpm of agitator. U.S. Varidrive motor, 2 HP, 3 phase, 60 cycle, 1800 rpm, explosion proof.

Vacuum:
- Three stage steam jets. 6 mm Hg absolute (0.058 psia).

Nitrogen:
- Pure, available at 20 psig.

Solution: The configuration for vessel charging is shown in Figure 3.17. The configuration for transferring the processed material from the reactor to the drums is shown in Figure 3.18. The complete procedure is shown in Appendix D and is formatted properly for operator use.

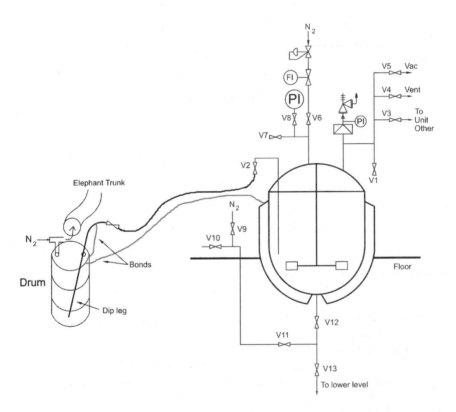

Figure 3.17. *Charging system for Example 3.2.*

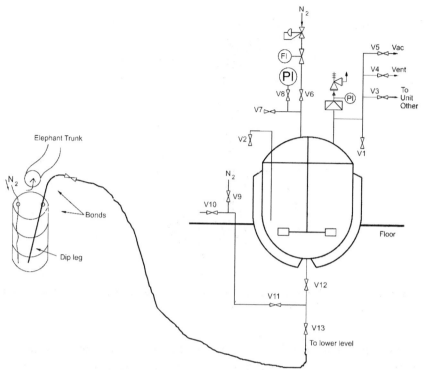

Figure 3.18. Drumming operation for Example 3.2.

3.12. Charging Powders

Several situations are possible here, depending on the combustibility of the powder and whether a flammable vapor is present. The flammable vapor may evolve from the powder, or may be present as a solvent in the process.

Static electricity formation can be expected during any bulk charging operation involving powders (Britton, 1999), even with gravity operations. If flammable vapors are absent, then static electricity formation does not normally present a problem if proper grounding and bonding are employed, no nonconductive materials are present, or large pieces of tramp metal are not present. Under these conditions static ignitions can only occur via bulking brush discharges during powder operations at high flow rates (see Section 2.14.1). Bulking brush discharges typically have a low probability of powder ignition in most cases. See Britton (1999) for more details on this issue.

Figure 3.19 shows the charging configuration for a powder being charged into a vessel without a flammable vapor present. The powder must have an MIE high enough that the powder will not ignite during the charging operation.

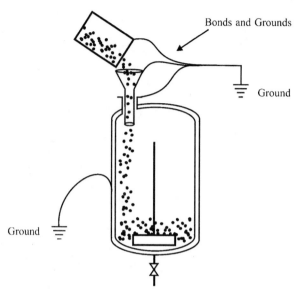

Figure 3.19. *Handling solids with no flammable vapors present (Expert Commission for Safety in the Swiss Chemical Industry, 1988).*

If flammable vapors are present, then the energy necessary for ignition is lower and, therefore, the probability of ignition increases greatly. Special handling is required for this case. First, inerting should be provided to reduce the oxygen concentration in the vessel so that the vapor space is not flammable. Second, a mechanical interlock must be provided. Two means of achieving this are shown in Figure 3.20. In the upper part of Figure 3.20, a rotary valve charges the powder as inert gas is supplied both to the rotary valve and to the vessel. In the lower part of Figure 3.20, the powder is first charged into a container, then the container is inerted with an inert gas prior to charging into the larger vessel.

Flexible intermediate bulk containers (FIBCs) have replaced many manual powder charging operations that previously used much smaller containers that a worker could lift. FIBCs are collapsible, square-shaped containers constructed of film, woven fabric or other material. They have a volume of less than 1.5 m^3 and bulk capacities of from 400 to 1000 kg of powder. They are typically lifted by a hoist over a dumping station and dumped by gravity through a fabric spout in the bottom of the FIBC. The entire contents can be dumped within a matter of minutes or less. The matter is complicated by static brush discharge within the bulk powder as the particles tumble over each other during the dumping operation. Furthermore, the electric fields produced during the dumping may induce static charges in adjacent, ungrounded objects.

Several types of FIBC are available: Types A, B, C, and D. Types A and B cannot be grounded, while type C is conductive and requires grounding and type D

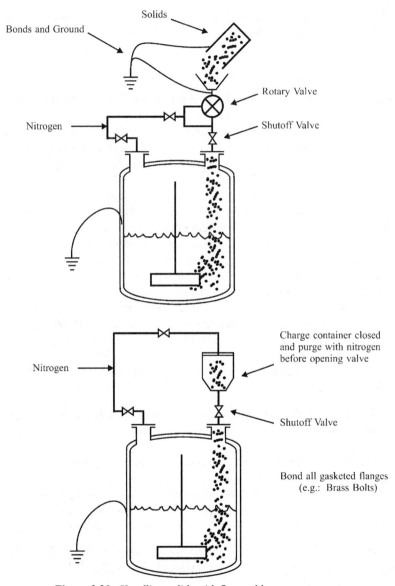

Figure 3.20. *Handling solids with flammable vapors present.*

is an antistatic type that does not require grounding. Type C contains a conductive grid and requires grounding. Care must be taken to ensure that the ground is in place and functioning during the entire powder charging operation. Type D incorporates a system of independent conductive fibers and is sufficiently conductive to dissipate charge.

Issues related to static generation and accumulation during FIBC loading and unloading are beyond the scope of this book. More details are provided elsewhere (Britton, 1993, 1999).

3.13. Electrical Equipment in Hazardous (Classified) Areas

The intent of providing special electrical equipment in hazardous or classified locations is to remove potential sources of ignition from an area that might contain flammable materials. As shown in Section 2.1.6, the minimum ignition energies are on the order of 0.25 mJ for flammable gases and about 10 mJ for dusts (although wide variability exists). Thus, very little energy is required to ignite gases and dusts—much less energy than is typically found in a common electrical circuits or a small static spark. Furthermore, hot surfaces of light bulbs and electrical motors represent additional potential ignition sources, since these surfaces might exceed the autoignition temperature of some chemical species.

Locations are classified depending on the properties of the flammable vapors, liquids or gases, or combustible dusts or fibers that may be present and the likelihood that a flammable or combustible concentration or quantity is present. Where pyrophoric materials are the only materials used or handled, these locations are not classified. Each room, section, or area is considered individually in determining its classification.

The classes are related to the nature of the flammable material (NFPA 30, 2002):

- *Class I:* Locations in which flammable gases or vapors are or may be present in the air in quantities sufficient to produce explosive or ignitible mixtures.
- *Class II:* Locations that are hazardous because of the presence of combustible dust.
- *Class III:* Locations that are hazardous because of the presence of easily ignitible fibers or flyings, but such fibers or flyings are not likely suspended in the air in sufficient quantities to produce ignitible mixtures.

Division designations are categorized based on the probability of the material being within the flammable or explosive regions (NFPA 30, 2002):

- *Division 1:* Probability of ignition is high. This means that ignitible concentrations exist under normal operating conditions, or exist frequently because of repair or maintenance operations or because of leakage, or breakdown or faulty operation of equipment or processes.
- *Division 2:* Hazardous only under abnormal conditions. Flammable materials are handled, processed, or used, but normally within closed containers or closed systems from which they escape only in case of accidental rupture

or breakdown of the containers or systems or abnormal operation of equipment. For this division, ignitible concentrations may be prevented by positive mechanical ventilation, and which might become hazardous through failure or abnormal operation of the ventilating equipment. Finally, this division is also used for locations adjacent to a Class I, Division 1 location from which ignitible materials might occasionally be carried by air flow, unless such movement is prevented by adequate positive-pressure ventilation from a source of clean air, and effective safeguards against ventilation failure are provided.

Group designations are also provided for specific types of chemicals or materials. Chemicals or materials within the same group are assumed to represent equivalent hazard. These groups are:

Class I:
- *Group A:* Acetylene
- *Group B:* Hydrogen, ethylene
- *Group C:* Carbon monoxide, hydrogen sulfide
- *Group D:* Butane, ethane, ethyl alcohol

Class II:
- *Group E:* Atmospheres containing combustible metal dusts, including aluminum, magnesium, and their commercial alloys, or other combustible dusts whose particle size, abrasiveness, and conductivity present similar hazards in the use of electrical equipment.
- *Group F:* Atmospheres containing combustible carbonaceous dusts that have more than 8 percent total entrapped volatiles. This includes carbon black.
- *Group G:* Atmospheres containing combustible dusts not included in Group E or F, including flour, grain, wood, plastic, and chemicals.

Groups B, C, and D are also defined in terms of the **maximum experimental safe gap** (MESG) and the **minimum igniting current** ratio (MIC). The MESG is the maximum clearance between two parallel metal surfaces that has been found, under specified test conditions, to prevent an explosion in a test chamber from being propagated to a secondary chamber containing the same gas or vapor at the same concentration. The MIC ratio is the ratio of the minimum current required from an inductive spark discharge to ignite the most easily ignitable mixture of a gas or vapor, divided by the minimum current required from an inductive spark discharge to ignite methane under the same test conditions. Table 3.4 provides the MESG and MIC ratio values for the various groups.

Table 3.5 provides examples of Class I, Division 1 locations.

In 1996 the National Electrical Code, NEC (NFPA 70, 2002) introduced Article 505 which is based primarily on the standards of the International Electrotechnical Commission (IEC). Article 505 provides definitions of hazardous

TABLE 3.4
MESGs and MIC Ratios for the Various Groups

Group	Examples	MESG	MIC Ratio
B	Hydrogen	≤0.45 mm	≤0.40
C	Carbon monoxide, hydrogen sulfide, ethylene	>0.45 mm ≤0.75 mm	>0.40 ≤0.80
D	Butane, ethane, ethyl alcohol	>0.75 mm	>0.80

TABLE 3.5
Examples of Class I, Division 1 locations from Section 500.5 of NFPA 70 (2001)[a]

- Locations where volatile flammable liquids or liquefied flammable gases are transferred from one container to another.
- Interiors of spray booths and areas in the vicinity of spraying and painting operations where volatile flammable solvents are used .
- Locations containing open tanks or vats of volatile flammable liquids.
- Drying rooms or compartments for the evaporation of flammable solvents.
- Locations containing fat- and oil-extraction equipment using volatile flammable solvents.
- Portions of cleaning and dyeing plants where flammable liquids are used.
- Gas generator rooms and other portions of gas manufacturing plants where flammable gas may escape.
- Inadequately ventilated pump rooms for flammable gas or for volatile flammable liquids.
- The interiors of refrigerators and freezers in which volatile flammable materials are stored in open, lightly stoppered, or easily ruptured containers.

In some Division 1 locations, ignitible concentrations of flammable gases or vapors may be present continuously or for long periods of time. Examples include the following:

- The inside of inadequately vented enclosures containing instruments normally venting flammable gases or vapors to the interior of the enclosure.
- The inside of vented tanks containing volatile flammable liquids.
- The area between the inner and outer roof sections of a floating roof tank containing volatile flammable fluids.
- Inadequately ventilated areas within spraying or coating operations using volatile flammable fluids.
- The interior of an exhaust duct that is used to vent ignitible concentrations of gases or vapors.

[a]See also NFPA 30 (2002), NFPA 58 (2001) and NFPA 497 (1997).

locations which are used internationally. At this time, Article 505 discusses only areas exposed to flammable gas and vapor atmospheres and does not deal with flammable or combustible dusts. In Article 505, Class I applies to locations where flammable gases or vapors may be present, identical to the Class I locations discussed previously. Furthermore, a number of zones are defined:

A Class I, Zone 0 location is one in which

1. ignitible concentrations of flammable gases or vapors are present continuously, or
2. ignitible concentrations of flammable gases or vapors are present for long periods of time.

A Class I, Zone 1 location is one

1. in which ignitible concentrations of flammable gases or vapors are likely to exist under normal operating conditions; or
2. in which ignitible concentrations of flammable gases or vapors may exist frequently because of repair or maintenance operations or because of leakage; or
3. in which equipment is operated or processes are carried on, of such a nature that equipment breakdown or faulty operations could result in the release of ignitible concentrations of flammable gases or vapors and also cause simultaneous failure of electrical equipment in a mode to cause the electrical equipment to become a source of ignition; or
4. that is adjacent to a Class I, Zone 0 location from which ignitible concentrations of vapors could be communicated, unless communication is prevented by adequate positive pressure ventilation from a source of clean air and effective safeguards against ventilation failure are provided.

A Class I, Zone 2 location is one

1. in which ignitible concentrations of flammable gases or vapors are not likely to occur in normal operation and, if they do occur, will exist only for a short period; or
2. in which volatile flammable liquids, flammable gases, or flammable vapors are handled, processed, or used but in which the liquids, gases or vapors normally are confined within closed containers of closed systems from which they can escape, only as a result of accidental rupture or breakdown of the containers or system, or as a result of the abnormal operation of the equipment with which the liquids or gases are handled, processed, or used; or
3. in which ignitible concentrations of flammable gases or vapors normally are prevented by positive mechanical ventilation but which may become hazardous as a result of failure or abnormal operation of the ventilation equipment; or

4. that is adjacent to a Class I, Zone 1 location, from which ignitible concentrations of flammable gases or vapors could be communicated, unless such communication is prevented by adequate positive-pressure ventilation from a source of clean air and effective safeguards against ventilation failure are provided.

The IEC system has three groups for vapors versus the four groups for vapors under the NEC system. These groups are translated as follows:

NEC Group	Equivalent IEC Group
A and B	IIC
C	IIB
D	IIA

A facility can choose to use either the Division or Zone system for electrical classification of an area. Equipment must be properly rated for the chosen Division Group or Zone Group. When different classification systems are used in contiguous areas, a Zone 2 area would abut a Division 2 area. More information is given in NFPA 497 (1997).

Table 3.6 provides information useful for determining the extent of the hazardous area in the vicinity of various types of equipment. For example, for indoor pumps in a Class I, Division 2 applications, the classified area extends within 5-feet of the pump in all directions, and also up to 3 feet above floor or grade level within 25 feet horizontally from any edge of the pump.

TABLE 3.6
Electrical Area Classifications for Flammable Liquids,
from Table 6.2.2 of NFPA 30 (2000)[a]

Hazardous Location	NEC Class I		Extent of Classified Area
	Division	Zone	
Indoor equipment installed where flammable vapor–air mixtures can exist under normal operation	1	0	The entire area associated with such equipment where flammable gases or vapors are present continuously or for long periods of time
	1	1	Area within 5 ft of any edge of such equipment, extending in all directions
	2	2	Area between 5 ft and 8 ft of any edge of such equipment, extending in all directions; also, space up to 3 ft above floor or grade level within 5 ft to 25 ft horizontally from any edge of such equipment

Hazardous Location	NEC Class I		Extent of Classified Area
	Division	Zone	
Outdoor equipment of the type where flammable vapor–air mixtures can exist under normal operation	1	0	The entire area associated with such equipment where flammable gases or vapors are present continuously or for long periods of time
	1	1	Area within 3 ft of any edge of such equipment, extending in all directions
	2	2	Area between 3 ft and 8 ft of any edge of such equipment, extending in all directions; also, space up to 3 ft above floor or grade level within 3 ft to 10 ft horizontally from any edge of such equipment
Tank storage installations inside buildings	1	1	All equipment located below grade level.
	2	2	Any equipment located at or above grade level.
Tank—underground	1	0	Inside fixed-roof tank
	1	1	Area inside dike where dike height is greater than the distance from the tank to the dike for more than 50 percent of the tank circumference
Shell, ends, or roof and dike area	2	2	Within 10 ft from shell, ends, or roof of tank; also, area inside dikes to level of top of tank
Vent	1	0	Area inside of vent piping or opening
	1	1	Within 5 ft of open end of vent, extending in all directions
	2	2	Area between 5 ft and 10 ft from open end of vent, extending in all directions
Floating roof, with fixed outer roof	1	0	Area between the floating and fixed-roof sections and within the shell
Floating roof, without fixed outer roof	1	1	Area above the floating roof and within the shell
Underground tank fill opening	1	1	Any pit, box, or space below grade level, if any part is within a Division 1 or 2 or Zone 1 or 2 classified location
	2	2	Up to 18 in. above grade level within a horizontal radius of 10 ft from a loose fill connection and within a horizontal radius of 5 ft from a tight fill connection

| Hazardous Location | NEC Class I | | Extent of Classified Area |
	Division	Zone	
Vent— discharging upward	1	0	Area inside of vent piping or opening
	1	1	Within 3 ft of open end of vent, extending in all directions
	2	2	Area between 3 ft and 5 ft of open end of vent, extending in all directions
Drum and container filling— outdoors or indoors	1	0	Area inside the drum or container
	1	1	Within 3 ft of vent and fill openings, extending in all directions
	2	2	Area between 3 ft and 5 ft from vent or fill opening, extending in all directions; also, up to 18 in. above floor or grade level within a horizontal radius of 10 ft from vent or fill opening.
Pumps, bleeders, withdrawal fittings—indoor	2	2	Within 5 ft of any edge of such devices, extending in all directions; also, up to 3 ft above floor or grade level within 25 ft horizontally from any edge of such devices.
Pumps, bleeders, withdrawal fittings—outdoor	2	2	Within 3 ft of any edge of such devices, extending in all directions; also, up to 18 in. above grade level within 10 ft horizontally from any edge of such devices.
Pits and sumps without mechanical ventilation	1	1	Entire area within a pit or sump if any part is within a Division 1 or 2 or Zone 1 or 2 classified location.
Pits and sumps with adequate mechanical ventilation	2	2	Entire area within a pit or sump if any part is within a Division 1 or 2 or Zone 1 or 2 classified location.
Containing valves, fittings, or piping, and not within a Division 1 or 2 or Zone 1 or 2 classified location	2	2	Entire pit or sump.
Drainage ditches, separators, impounding basins—outdoor	2	2	Area up to 18 in. above ditch, separator, or basin; also, area up to 18 in. above grade within 15 ft horizontally from any edge
Drainage ditches, separators, impounding basins—indoor			Same classified area as pits.
Tank vehicle and tank car loading through open dome	1	0	Area inside of the tank.
	1	1	Within 3 ft of edge of dome, extending in all directions
	2	2	Area between 3 ft and 15 ft from edge of dome, extending in all directions

| Hazardous Location | NEC Class I | | Extent of Classified Area |
	Division	Zone	
Loading through bottom connections with atmospheric venting	1	0	Area inside of the tank.
	1	1	Within 3 ft of point of venting to atmosphere, extending in all directions.
	2	2	Area between 3 ft and 15 ft from point of venting to atmosphere, extending in all directions; also, up to 18 in. above grade within a horizontal radius of 10 ft from point of loading connection
Office and rest rooms	Ordi-nary	Ordi-nary	If there is any opening to these rooms within the extent of an indoor classified location, the room shall be classified the same as if the wall, curb, or partition did not exist
Loading through closed dome with atmospheric venting	1	1	Within 3 ft of open end of vent, extending in all directions.
	2	2	Area between 3 ft and 15 ft from open end of vent, extending in all directions; also, within 3 ft of edge of dome, extending in all directions.
Loading through closed dome with vapor control	2	2	Within 3 ft of point of connection of both fill and vapor lines, extending in all directions.
Bottom loading with vapor control or any bottom unloading	2	2	Within 3 ft of point of connections, extending in all directions; also, up to 18 in. above grade within a horizontal radius of 10 ft from point of connections
Storage and repair garage for tank vehicles	1	1	All pits or spaces below floor level.
	2	2	Area up to 18 in. above floor or grade level for entire storage or repair garage
Garages for other than tank vehicles	Ordi-nary	Ordi-nary	If there is any opening to these rooms within the extent of an outdoor classified location, the entire room shall be classified the same as the area classification at the point of the opening.
Outdoor drum storage	Ordi-nary	Ordi-nary	
Inside rooms or storage lockers used for the storage of Class I liquids	2	2	Entire room.
Indoor warehousing where there is no flammable liquid transfer	Ordi-nary	Ordi-nary	If there is any opening to these rooms within the extent of an indoor classified location, the room shall be classified the same as the wall, curb, or partition did not exist.
Piers and wharves			See NFPA 30.

[a]Used by permission of the National Fire Protection Association.

3.13.1. Protection Techniques

Table 3.7 summarizes the protection methods described by the NEC (NFPA 70, 2002).

TABLE 3.7
Various Types of Electrical Protection Methods and the Applicable Areas[a]

Protection Method	Definition	Location
Dust ignition proof	Equipment enclosed in a manner that excludes dusts and does not permit arcs, sparks, or heat otherwise generated or liberated inside of the enclosure to cause ignition of exterior accumulations or atmospheric suspensions of a specified dust on or in the vicinity of the enclosure.	Class II, Division 1 or 2
Dust tight	Enclosures constructed so that dust will not enter under specified test conditions.	Class II, Division 2 or Class III, Division 1 or 2
Explosion proof apparatus	Apparatus enclosed in a case that is capable of withstanding an explosion of a specified gas or vapor that may occur within it and of preventing the ignition of a specified gas or vapor surrounding the enclosure by sparks, flashes, or explosion of the gas or vapor within, and that operates at such an external temperature that a surrounding flammable atmosphere will not be ignited thereby.	Class I, Division 1 or 2
Hermetically sealed	Equipment sealed against the entrance of an external atmosphere where the seal is made by fusion, for example, soldering, brazing, welding, or the fusion of glass to metal.	Class I, Division 2; Class II, Division 2; or Class III, Division 1 or 2
Nonincendive circuit	A circuit, other than field wiring, in which any arc or thermal effect produced under intended operating conditions of the equipment is not capable, under specified test conditions, of igniting the flammable gas–air, vapor–air, or dust–air mixture.	Class I, Division 2; Class II, Division 2; or Class III, Division 1 or 2
Nonincendive component.	A component having contacts for making or breaking an incendive circuit and the contacting mechanism is constructed so that the component is incapable of igniting the specified flammable gas–air or vapor–air mixture. The housing of a nonincendive component is not intended to exclude the flammable atmosphere or contain an explosion.	Class I, Division 2; Class II, Division 2; or Class III, Division 1 or 2

Protection Method	Definition	Location
Nonincendive equipment	Equipment having electrical/electronic circuitry that is incapable, under normal operating conditions, of causing ignition of a specified flammable gas–air, vapor–air, or dust–air mixture due to arcing or thermal means.	Class I, Division 2; Class II, Division 2; or Class III, Division 1 or 2
Nonincendive field wiring	Wiring that enters or leaves an equipment enclosure and, under normal operating conditions of the equipment, is not capable, due to arcing or thermal effects, of igniting the flammable gas–air, vapor–air, or dust–air mixture. Normal operation includes opening, shorting, or grounding the field wiring.	
Nonincendive field wiring apparatus	Apparatus intended to be connected to nonincendive field wiring.	
Oil immersion	Electrical equipment immersed in a protective liquid in such a way that an explosive atmosphere that may be above the liquid or outside the enclosure cannot be ignited.	Current interrupting contacts in Class I, Division 2
Purged and pressurized	The process of supplying an enclosure with a protective gas at a sufficient flow and positive pressure to reduce the concentration of any flammable gas or vapor initially present to an acceptable level.	Any hazardous (classified) location

[a]From Section 500.5 of NFPA 70 (2002).

The primary protection mechanism for electrical equipment operated in classified areas is by **explosion proof enclosures**. These enclosures are used to house any potential ignition sources, such as a switch, relay, or contactor, which have arcing contacts.

Electrical equipment that will be operated in an NEC Class I, Division 1 or Zone 1 area, is designed assuming that the enclosure could be filled with a flammable gas or vapor at its most optimum concentration and ignited. The enclosure must meet three requirements. The first requirement is that the enclosure must be capable of withstanding the internal pressure from the explosion. A safety factor of four is used, i.e. the enclosure must withstand a hydrostatic pressure equivalent to four times the maximum pressure during the combustion. Second, the expanding gas escaping through the flanges or screwed threads on the enclosure must be cooled sufficiently so that it will not ignite a flammable mixture at its most easily ignited state outside of the enclosure. The joints, flanges and threads of the enclosure are held within narrow tolerances so that the escaping hot gases from the internal combustion are cooled as they escape, preventing ignition of the external gases. This is shown in Figure 3.21. Third, the outer surface of the enclosure must

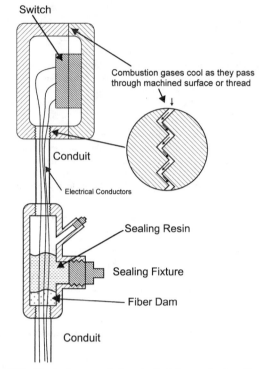

Figure 3.21. *Explosion proof electrical enclosure and sealing fixture.*

operate at a temperature cool enough to prevent autoignition of the surrounding gases.

It is a common misconception that the purpose of explosion proof enclosures is to prevent flammable gases from entering the fixture. This is not the case. The design is based on containing the internal explosion. Furthermore, another common misconception is that explosion proof fixtures are waterproof and can withstand rain, water and steam cleaning. This is also not true, and provision must be made to prevent water accumulation within the fixture.

The integrity of the quenching surfaces shown in Figure 3.21 is important to the operation of the enclosure. The flat, machined surfaces must not be scratched or dirty to prevent proper gas quenching. Electrical fixtures are frequently painted and a screwdriver is frequently used to separate the flanges to open the enclosure. This may cause a scratch on the machined surface, preventing proper quenching of the gases and defeating the design of the enclosure. The enclosure cover is also frequently set on the floor, enabling dirt to adhere to the machined surface and preventing proper operation of the enclosure.

It is also common practice to refer to areas requiring classified electrical equipment as "explosion proof" areas. This is a misuse of the term—only the equipment is explosion proof and not the area.

For Class II locations, the enclosure must keep the dust out of the interior and must operate at a safe surface temperature. In this case no internal explosion occurs and the device may have thinner walls than the Class I enclosure. Dust layering is a problem with these enclosures—dust build-up may insulate the fixture and lead to higher enclosure temperatures and potential ignition of the dust layer. Common terms used to define these fixtures are dust-ignition proof or dust tight, as defined in Table 3.7.

Another electrical equipment protection method is purging and pressurization, as defined in Table 3.7. Pressurized enclosures require (1) a source of clean air or inert gas, (2) a compressor to maintain the required pressure, (3) a pressure control valve to prevent the power from being applied before the enclosure has been purged, and to deenergize the system if the pressure falls below a safe value (Crouse Hinds, 1999). Where the enclosure is in a Class I, Division 1 or Zone 1, area and the electrical equipment inside is unclassified, an additional door interlock is required to de-energize the electrical equipment if the door is opened (NFPA 496, 1998). Purging of enclosures is an active system requiring frequent inspection, testing and maintenance to ensure the required protection. Purging is primarily used for large enclosures which would be prohibitively expensive as explosion proof housings.

Electrical motors and generators used in hazardous locations must also be properly designed. In Class I, Division 1 locations the only motors that may be used are either (1) explosion proof, (2) totally enclosed and pressurized with clean air, (3) totally enclosed and filled with inert gas, or (4) special submerged typed motors. Motors used in Class I, Division 2 locations containing sliding contacts or switching mechanisms must also be explosion proof or pressurized. Open type 3-phase motors such as squirrel-cage induction motors without any arcing devices may be used in Class I, Division 2 locations. Maintenance of rated motors and generators must be performed only by authorized personnel or contractors (Crouse Hinds, 1999).

Plugs and receptacles must also be specially designed for use in hazardous locations. In plugs and receptacles, the electrical contact is made outside of the enclosure, presenting an ignition hazard. Two design methods are used to prevent arcing and ignition. In the interlocked, dead front design, the receptacle contacts are interlocked with a switch located in an explosion proof enclosure. The receptacle contacts are not live when the plug is inserted or withdrawn. In the delayed action design, the plug and receptacle are designed so that the contacts are confined within an explosion proof enclosure. This design also prevents the rapid withdrawal of the plug, allowing any heated metal parts to cool before they come in contact with the surrounding location (Crouse Hinds, 1999).

Conduits and wiring must be designed specifically for hazardous locations (NFPA 70, 2002). All conduit must be of rigid metal with at least five full tapered threads tightly engaged in any enclosure. This thread requirement is deeper than standard NPT (National Pipe Thread). Sealing fittings are also required in conduits, as shown in Figure 3.21. The seals are necessary to (1) limit volume, (2) prevent an explosion from traveling down the conduit, (3) block gases from moving from a hazardous to a nonhazardous location, and (4) prevent pressure piling as explained in Section 2.7.4. Sealing fixtures are required in the following instances: (1) where the conduit enters an enclosure that houses arcing or high temperature equipment. In this case the sealing fixture must be within 18-inches of the enclosure. (2) where the conduit leaves a Division 1 area or passes from a Division 2 area to a nonhazardous location, (3) where the conduit enters enclosures that house terminals, splices or taps, but only if the conduit is 2-inches or more in diameter (Crouse Hinds, 1999).

The sealing is accomplished using an epoxy type sealing compound and a special sealing fixture. The wires in the fixture must first be separated. A fiber dam is then placed in the fixture to hold the liquid sealing compound. Finally, the liquid sealing compound is carefully placed into the fitting, behind the dam and between the separated wires. The sealing compound eventually hardens, providing the proper seal. Most chemical plants with hazardous locations may contain hundreds, if not thousands, of sealing fixtures. A common problem is ensuring that all the fixtures have been sealed. One method employed is to spray paint the fixture with a special color after the sealing has been completed. Accidents have occurred due to unsealed or improperly sealed sealing fixtures, allowing flammable vapor to travel down a conduit and reaching an arcing ignition source.

Intrinsically safe or nonincendive equipment is primarily used for process control instrumentation since these systems lend themselves to low energy requirements. The purpose is to provide electrical current at levels less than the minimum ignition energies (MIE) under normal and abnormal conditions. Additional details are provided in the National Electrical Code (NEC) (NFPA 70, 2002).

Appendix A

DETAILED EQUATIONS FOR FLAMMABILITY DIAGRAMS

Part A: Equations Useful for Gas Mixtures

This appendix derives several equations that are useful for working with flammability diagrams. Section 2.1.1 provides introductory material on the flammability diagram. This section will derive equations proving that:

1. If two gas mixtures, R and S, are combined, the resulting mixture composition lies on a line connecting the points R and S on the flammability diagram. The location of the final mixture on the straight line depends on the relative moles of the mixtures combined—if mixture S has more moles, the final mixture point will lie closer to point S. This is identical to the lever rule, which is used for phase diagrams.
2. If a mixture R is continuously diluted with mixture S, the mixture composition will follow along the straight line between points R and S on the flammability diagram. As the dilution continues, the mixture composition will move closer and closer to point S. Eventually, at infinite dilution, the mixture composition will be at point S.
3. For systems having composition points that fall on a straight line passing through an apex corresponding to one pure component, the other two components are present in a fixed ratio along the entire line length.
4. The limiting oxygen concentration (LOC) is estimated by reading the oxygen concentration at the intersection of the stoichiometric line and a horizontal line drawn through the LFL. This is equivalent to the equation

$$LOC = z(LFL) \qquad (A-1)$$

Figure A.1 shows two gas mixtures, denoted by R and S, that are combined to form mixture M. Each gas mixture has a specific composition, based on the three gas components, A, B, and C. For mixture R, the gas composition, in mole fractions, is x_{AR}, x_{BR}, and x_{CR}, and the total number of moles is n_R. For mixture S the

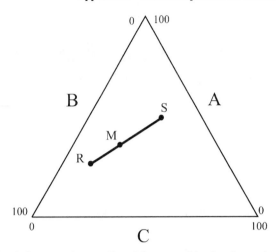

Figure A.1. *Two mixtures, R and S, are combined to form mixture M.*

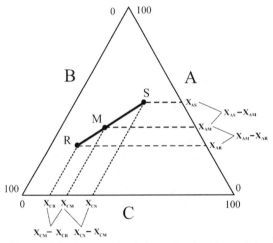

Figure A.2. *Composition information for Figure A.1.*

gas composition is x_{AS}, x_{BS}, and x_{CS}, with total moles n_S, and at mixture M the gas composition is x_{AM}, x_{BM}, and x_{CM}, with total moles n_M. These compositions are shown on Figure A.2 with respect to components A and C.

An overall and a component species balance can be performed to represent the mixing process. Since a reaction does not occur during mixing, moles are conserved and it follows that,

$$n_M = n_R + n_S \qquad\qquad (A\text{-}2)$$

A mole balance on species A is given by

$$n_M x_{AM} = n_R x_{AR} + n_S x_{AS} \qquad \text{(A-3)}$$

A mole balance on species C is given by

$$n_M x_{CM} = n_R x_{CR} + n_S x_{CS} \qquad \text{(A-4)}$$

Substituting Equation (A-2) into Equation (A-3) and rearranging,

$$\frac{n_S}{n_R} = \frac{x_{AM} - x_{AR}}{x_{AS} - x_{AM}} \qquad \text{(A-5)}$$

Similarly, substituting Equation (A-2) into Equation (A-4) results in,

$$\frac{n_S}{n_R} = \frac{x_{CM} - x_{CR}}{x_{CS} - x_{CM}} \qquad \text{(A-6)}$$

Equating Equations (A-5) and (A-6) results in,

$$\frac{x_{AM} - x_{AR}}{x_{AS} - x_{AM}} = \frac{x_{CM} - x_{CR}}{x_{CS} - x_{CM}} \qquad \text{(A-7)}$$

A similar set of equations can be written between components A and B, or B and C.

Figure A.2 shows the quantities represented by the mole balance of Equation (A-7). The mole balance, Equation (A-7), is honored only if point M lies on the straight line between points R and S. This can be shown on Figure A.2 using similar triangles (Hougen, Watson et al., 1954).

Figure A.3 shows another useful result based on Equations (A-5) and (A-6). These equations imply that the location of point M on the straight line between points R and S depend on the relative moles of R and S, as shown.

The above result can, in general, be applied to any two points on the triangle diagram. If a mixture R is continuously diluted with mixture S, the mixture composition will follow along the straight line between points R and S. As the dilution continues, the mixture composition will move closer and closer to point S. Eventually, at infinite dilution, the mixture composition will be at point S.

For systems having composition points that fall on a straight line passing through an apex corresponding to one pure component, the other two components are present in a fixed ratio along the entire line length (Hougen, Watson et al., 1954). This is shown in Figure A.4. For this case the ratio of components A and B along the line shown is constant and is given by

$$\frac{x_A}{x_B} = \frac{x}{100 - x} \qquad \text{(A-8)}$$

A useful application of this result is shown on Figure A.5. Suppose we wish to find the oxygen concentration at the point where the LFL intersects the

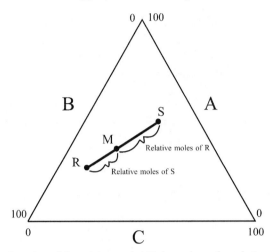

Figure A.3. *The location of the mixture point M depends on the relative masses of mixtures R and S.*

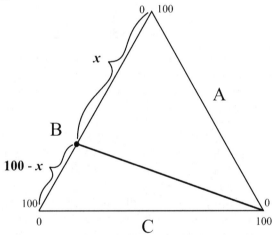

Figure A.4. *The ratio of components A and B is constant along the line shown and is given by x/(100 − x).*

stoichiometric line shown. The oxygen concentration in question is shown as point x on Figure A.5. The stoichiometric combustion equation is represented by

$$(1) \text{ Fuel} + z \text{ Oxygen} \rightarrow \text{Products} \tag{A-9}$$

where z is the stoichiometric coefficient for oxygen. The ratio of oxygen to fuel along the stoichiometric line is constant and is given by

$$\frac{x_{O_2}}{x_{\text{fuel}}} = z \tag{A-10}$$

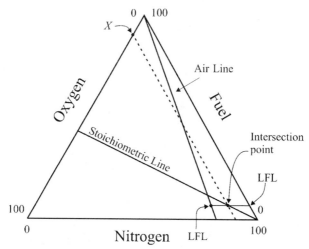

Figure A.5. *Determining the oxygen concentration, X, at the intersection of the LFL and the stoichiometric line.*

At the specific fuel concentration of x_{fuel} = LFL, it follows from Equation (A-10) that

$$x_{O_2} = z(LFL) \qquad (A\text{-}11)$$

This result provides a method to estimate the limiting oxygen concentration (LOC) from the LFL—the nose of the flammable envelope is usually very close to the intersection of a horizontal line drawn through the LFL and the stoichiometric line. For many hydrocarbons the nose is slightly above the stoichiometric line—in this case the estimating method of Equation (A-11) provides a conservative result. This graphical estimate of the LOC is equivalent to the method provided by Crowl and Louvar (Crowl and Louvar, 2002) using the lower flammability limit. In their approach, for a combustion reaction given by Equation (A-9), the LOC is estimated by:

$$LOC = z \, (LFL) \qquad (A\text{-}12)$$

where z is the stoichiometric coefficient for oxygen, given by Equation (A-9) and LFL is the lower flammability limit, in volume percent fuel in air.

Part B: Equations Useful for Placing Vessels Into and Out of Service

The equations presented in this section are equivalent to drawing straight lines to show the gas composition transitions, as shown in Section 3.1.1. The equations are

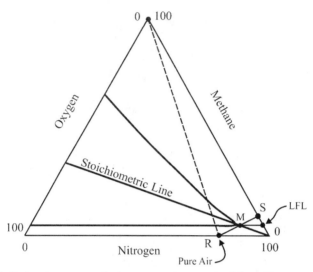

Figure A-6. *Estimating a target fuel concentration at point S for taking a vessel out of service.*

frequently easier to use and provide a more precise result than manually drawn lines.

The out-of-service fuel concentration (OSFC) represents the maximum fuel concentration that just avoids the flammability zone when a vessel is being taken out of service. It is shown as point S on Figure A.6.

For most compounds detailed flammability zone data are not available. In this case an estimate can be made of the location of point S, as shown in Figure A.6. Point S can be approximated by a line starting at the pure air point and connecting through a point at the intersection of the LFL with the stoichiometric line. Equation (A-7) can be used to determine the gas composition at point S. Referring to Figure A.2, we know the gas composition at point R and M, and wish to calculate the gas composition at point S. Let A represent the fuel and C the oxygen, then from Figure A.2 and A.6 it follows that $x_{AR} = 0$, $x_{AM} = LFL\%$, x_{AS} is the unknown OSFC, $x_{CM} = z * LFL$ from Equation (A-11), $x_{CR} = 21\%$, and $x_{CS} = 0$. Then, by substitution into Equation (A-7) and solving for x_{AS} we get

$$x_{AS} = OSFC = \frac{LFL\%}{1 - z(LFL\%/21)} \tag{A-13}$$

where OSFC is the out-of-service fuel concentration, that is, the fuel concentration at point S on Figure A.6; LFL% is the volume percent of fuel in air at the lower flammability limit; and z is the stoichiometric oxygen coefficient from the combustion reaction given by Equation (A-9).

Another approach is to estimate the fuel concentration at point S by extending the line from point R through the intersection of the limiting oxygen concentration (LOC) and the stoichiometric line. The result is

$$OSFC = \frac{LOC\%}{z(1 - LOC\%/21)} \tag{A-14}$$

where LOC% is the minimum oxygen concentration in volume percent oxygen.

Equations (A-13) and (A-14) are approximations to the fuel concentration at point S. Fortunately, they are usually conservative, predicting a fuel concentration that is less than the experimentally determined OSFC value. For instance, for methane the LFL is 5.3% (Appendix C) and z is 2. Thus, Equation (A-13) predicts an OSFC of 10.7% fuel. This is compared to the experimentally determined OSFC of 14.5% (Table 3-1). Using the experimental LOC of 12% (Table 2.2), an OSFC value of 14% is determined using Equation (A-14). This is closer to the experimental value, but still conservative. For ethylene, 1,3-butadiene, and hydrogen, Equation (A-14) predicts a higher OSFC than the experimentally determined value. For all other species in Table 3.1, Equation (A-14) estimates an OSFC that is less than the experimental value.

The in-service oxygen concentration (ISOC) represents the maximum oxygen concentration that just avoids the flammability zone, shown as point S on Figure A.7. One approach to estimate the ISOC is to use the intersection of the LFL with the stoichiometric line. A line is drawn from the top apex of the triangle, through the intersection to the nitrogen axis, as shown on Figure A.7. Let A represent the fuel species and C the oxygen. Then, from Figure A.7 it follows that $x_{AM} = LFL\%$,

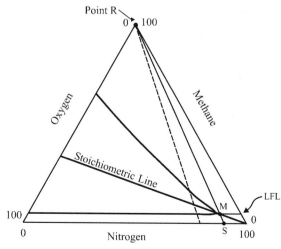

Figure A.7. Estimating a target nitrogen concentration at point S for placing a vessel into service.

$x_{AR} = 100$, $x_{AS} = 0$, $x_{CM} = z * LFL\%$ from Equation (A-11), $x_{CR} = 0$ and x_{CS} is the unknown ISOC. Substituting into Equation (A-7) and solving for the ISOC results in,

$$ISOC = \frac{z * LFL\%}{1 - (LFL\%/100)} \tag{A-15}$$

where ISOC is the in-service oxygen concentration in volume % oxygen; z is the stoichiometric coefficient for oxygen given by Equation (A-9); and LFL% is the fuel concentration at the lower flammability limit, in volume percent fuel in air.

The nitrogen concentration at point S is equal to $100 - ISOC$.

An expression to estimate the ISOC using the intersection of the minimum oxygen concentration and the stoichiometric line can also be developed using a similar procedure. The result is

$$ISOC = \frac{z * LOC\%}{z - (LOC\%/100)} \tag{A-16}$$

where LOC% is the limiting oxygen concentration in volume percent oxygen.

Comparison of the estimates using Equations (A-15) and (A-16) to the experimental values in Table 3-2 shows that Equation (A-15) predicts a lower oxygen value than the experimental values for all species, with the exception of methyl formate. Equation (A-16) predicts a lower oxygen concentration than the experimental value for all species in Table 3-1 with the exception of butane, 3-methyl-1-butene, 1,3-butadiene, isobutyl formate, and acetone.

Direct, reliable experimental data under conditions as close as possible to process conditions is always recommended.

Appendix B

EQUATIONS FOR DETERMINING THE ENERGY OF EXPLOSION

Four methods are used to estimate the energy of explosion for a pressurized gas: Brode's equation, isentropic expansion, isothermal expansion, and thermodynamic availability.

Brode's method (Brode, 1959) is perhaps the simplest approach. It determines the energy required to raise the pressure of the gas at constant volume from ambient pressure to the burst pressure of the vessel. The resulting expression is

$$E = \frac{(P_2 - P_1)V}{\gamma - 1} \qquad \text{(B-1)}$$

where E is the energy of explosion (energy)

P_1 is the ambient pressure (force/area)

P_2 is the burst pressure of the vessel (force/area)

V is the volume of expanding gas in the vessel (volume)

γ is the heat capacity ratio for the gas (unitless)

Since $P_2 > P_1$, the energy calculated by Equation (B-1) is positive, indicating that the energy is released to the surroundings during the vessel rupture.

The isentropic expansion method assumes the gas expands isentropically from its initial to final state. The following equation represents this case (Smith and Ness, 1987),

$$E = \left(\frac{P_2 V}{\gamma - 1} \right) \left[1 - \left(\frac{P_1}{P_2} \right)^{(\gamma-1)/\gamma} \right] \qquad \text{(B-2)}$$

The isothermal case assumes that the gas expands isothermally. This is represented by the following equation (Smith and Ness, 1987),

$$E = R_g T_1 \ln\left(\frac{P_2}{P_1} \right) = P_2 V \ln\left(\frac{P_2}{P_1} \right) \qquad \text{(B-3)}$$

where R_g is the ideal gas constant and T_1 is the ambient temperature (deg).

The final method uses thermodynamic availability to estimate the energy of explosion. Thermodynamic availability represents the maximum mechanical energy extractable from a material as it moves into equilibrium with the environment. The resulting overpressure from an explosion is a form of mechanical energy. Thus, thermodynamic availability predicts a maximum upper bound to the mechanical energy available to produce an overpressure.

An analysis by Crowl (1992) using batch thermodynamic availability resulted in the following expression to predict the maximum explosion energy of a gas contained within a vessel.

$$E = P_2 V \left[\ln \left(\frac{P_2}{P_1} \right) - \left(1 - \frac{P_1}{P_2} \right) \right] \tag{B-4}$$

Note that Equation (B-4) is nearly the same as Equation (B-3) for an isothermal expansion with the addition of a correction term. This correction term accounts for the energy lost as a result of the second law of thermodynamics.

The question arises as to which method to use. Figure B.1 presents the energy of explosion using all four methods as a function of initial gas pressure in the vessel. The calculation assumes an inert gas initially at 298 K with $\gamma = 1.4$. The gas expands into ambient air at 1 atm pressure. The isentropic method produces a low value for the energy of explosion. The isentropic expansion will result in a gas at a very low temperature—the expansion of an ideal gas from 200 psia to 14.7 psia results in a final temperature of 254°R, or −205°F. This is thermodynamically inconsistent since the final temperature is ambient. The isothermal method predicts a very large value for the energy of explosion because it assumes that all of the energy of compression is available to perform work. In reality, some of the

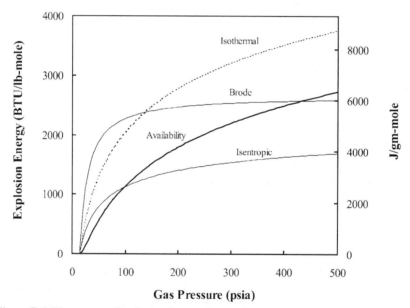

Figure B.1. *The energy of explosion for a compressed inert gas computed using the four available methods.*

energy must be expelled as waste heat according to the second law of thermody-namics. The availability method accounts for this loss by the correction term in Equation (B-4). All four methods continue to be used to estimate the energy of explosion for compressed gases.

It is thought that the Brode equation more closely predicts the potential explo-sion energy close to the explosion source, or near field, and that the isentropic expansion method predicts better the effects at a greater distance, or far field. How-ever, it is unclear where this transition occurs. Also, a portion of the potential explosion energy of vessel burst is converted into kinetic energy of the vessel pieces, and other inefficiencies (such as strain energy in the form of heat in the vessel fragments). For estimation purposes, it is not uncommon to subtract 50% of the total potential energy to calculate the blast pressure effects from vessel burst.

Some empirically based methods to estimate fragment formation and frag-ment range are based on the use of a specific energy equation above (typically the Brode equation). Since these methods are empirical, their use mandates the energy equation specified.

More detailed methods are available to calculate the energy of explosion for a rupturing vessel (Baker, Cox et al., 1988; AIChE, 1989, 1999a).

B.1. Example Application

EXAMPLE B.1
Determine the energy of explosion by the four methods for a 1 m^3 vessel contain-ing nitrogen at 500 bar abs pressure. The ambient pressure is 1.01 bar abs and the temperature is 298 K. Assume the vessel contains nitrogen with a constant heat capacity ratio of $\gamma = 1.4$.

Solution
For this case, $P_1 = 1.01$ bar and $P_2 = 500$ bar.

a. For Brode's method, Equation (B-1) is used

$$E = \frac{(P_2 - P_1)V}{\gamma - 1} = \frac{(500 \text{ bar} - 1.01 \text{ bar})(1 \text{ m}^3)}{1.4 - 1} = 1250 \text{ bar m}^3 = 1.25 \times 10^8 \text{ J}$$

b. For the isentropic method, Equation (B-2) is used

$$E = \left(\frac{P_2 V}{\gamma - 1} \right) \left[1 - \left(\frac{P_1}{P_2} \right)^{(\gamma-1)/\gamma} \right]$$

$$= \frac{(500 \text{ bar})(1 \text{ m}^3)}{1.4 - 1} \left[1 - \left(\frac{1.01 \text{ bar}}{500 \text{ bar}} \right)^{(1.4-1)/1.4} \right] = 1040 \text{ bar m}^3 = 1.04 \times 10^8 \text{ J}$$

c. For the isothermal method, Equation (B-3) is used

$$E = P_2 V \ln\left(\frac{P_2}{P_1}\right) = (500 \text{ bar})(1 \text{ m}^3) \ln\left(\frac{500 \text{ bar}}{1.01 \text{ bar}}\right)$$

$$= 3100 \text{ bar m}^3 = 3.10 \times 10^8 \text{ J}$$

d. For the availability method, Equation (B-4) is used

$$E = P_2 V \left[\ln\left(\frac{P_2}{P_1}\right) - \left(1 - \frac{P_1}{P_2}\right)\right]$$

$$= (500 \text{ bar})(1 \text{ m}^3)\left[\ln\left(\frac{500 \text{ bar}}{1.01 \text{ bar}}\right) - \left(1 - \frac{1.01 \text{ bar}}{500 \text{ bar}}\right)\right] = 2600 \text{ bar m}^3$$

$$= 2.60 \times 10^8 \text{ J}$$

The following table summarizes the results

Method	Energy (10^8 J)
Brode:	1.25
Isentropic:	1.04
Isothermal:	3.10
Availability:	2.60

As expected, the isentropic method produces the lowest value, while the isothermal method produces the largest value.

Appendix C

FLAMMABILITY DATA FOR SELECTED MATERIALS*

Compound	Formula	Heat of Combustion kJ/mol	Flammability Limit Vol. % Fuel in air		Flashpoint Temp.°C	Autoignition Temp. °C
			LFL	UFL		
Paraffin hydrocarbons						
Methane	CH_4	−890.3	5.3	15.0	−222.5	632
Ethane	C_2H_6	−1559.8	3.0	12.5	−130.0	472
Propane	C_3H_8	−2219.9	2.2	9.5	−104.4	493
Butane	C_4H_{10}	−2877.5	1.9	8.5	−60.0	408
Isobutane	C_4H_{10}	−2869.0	1.8	8.4	—	462
Pentane	C_5H_{12}	−3536.6	1.5	7.8	<−40.0	579
Isopentane	C_5H_{12}	−3527.6	1.4	7.6	—	420
2,2-Dimethylpropane	C_5H_{12}	−3514.1	1.4	7.5	—	450
Hexane	C_6H_{14}	−4194.5	1.2	7.5	−23.0	487
Heptane	C_7H_{16}	−4780.6	1.2	6.7	−4.0	451
2,3-Dimethylpentane	C_7H_{16}	−4842.3	1.1	6.7	—	337
Octane	C_8H_{18}	−5511.6	1.0	6.7	13.3	458
Nonane	C_9H_{20}	—	0.8	—	31.1	285
Decane	$C_{10}H_{22}$	−6737.0	0.8	5.4	46.1	463
Olefins						
Ethylene	C_2H_4	−1411.2	3.1	32.0	—	490
Propylene	C_3H_6	−2057.3	2.4	10.3	−107.8	458
1-Butene	C_4H_8	−2716.8	1.6	9.3	−80.0	384
2-Butene	C_4H_8	−2708.2	1.8	9.7	−73.3	435
1-Pentene	C_5H_{10}	−3361.4	1.5	8.7	−17.8	273

Compound	Formula	Heat of Combustion kJ/mol	Flammability Limit Vol. % Fuel in air		Flashpoint Temp.°C	Autoignition Temp. °C
			LFL	UFL		
Acetylenes						
Acetylene	C_2H_2	−1299.6	2.5	80.0	−17.8	305
Aromatics						
Benzene	C_6H_6	−3301.4	1.4	7.1	−11.1	740
Toluene	C_7H_8	−3947.9	1.4	6.7	4.4	810
o-Xylene	C_8H_{10}	−4567.6	1.0	6.0	17.0	496
Cyclic hydrocarbons						
Cyclopropane	C_3H_6	−2091.3	2.4	10.4	—	498
Cyclohexane	C_6H_{12}	−3953.0	1.3	8.0	−17.0	259
Methylcyclohexane	C_7H_{14}	−4600.7	1.2	—	—	265
Terpenes						
Turpentine	$C_{10}H_{16}$	—	0.8	—	35.0	252
Alcohols						
Methyl alcohol	CH_4O	−764.0	7.3	36.0	12.2	574
Ethyl alcohol	C_2H_6O	−1409.2	4.3	19.0	12.8	558
2-propen-1-ol	C_3H_6O	−1912.2	2.5	18.0	21.1	389
n-Propyl alcohol	C_3H_8O	−2068.9	2.1	13.5	15.0	505
Isopropyl alcohol	C_3H_8O	−2051.0	2.0	12.0	11.7	590
n-Butyl alcohol	$C_4H_{10}O$	−2728.3	1.4	11.2	35.0	450
Amyl alcohol	$C_5H_{12}O$	−3320.8	1.2	—	32.8	409
Isoamyl alcohol	$C_5H_{12}O$	—	1.2	—		518
Aldehydes						
Acetaldehyde	C_2H_4O	−764.0	4.1	57.0	−37.8	185
Crotonic aldehyde	C_4H_6O	−2268.1	2.1	15.5	12.8	—
2-Furancarboxaldehyde	$C_5H_4O_2$	−2340.9	2.1	—	—	—
Paraldehyde	$C_6H_{12}O_3$	—	1.3	—	17.0	541
Ethers						
Diethyl ether	$C_4H_{10}O$	−2751.1	1.9	48.0	−45.0	229
Divinyl ether	C_4H_6O	−2416.2	1.7	27.0	<−30.0	360

Compound	Formula	Heat of Combustion kJ/mol	Flammability Limit Vol. % Fuel in air		Flashpoint Temp.°C	Autoignition Temp. °C
			LFL	UFL		
Ketones						
Acetone	C_3H_6O	−1821.4	3.0	13.0	−17.8	700
Methylethyl ketone	C_4H_8O	−2478.7	1.8	10.0	−4.4	514
Methylpropyl ketone	$C_5H_{10}O$	−3137.6	1.5	8.0	7.2	505
Methylbutyl ketone	$C_6H_{12}O$	−3796.3	1.3	8.0	—	533
Acids						
Acetic acid	$C_2H_4O_2$	−926.1	5.4	—	42.8	599
Hydrocyanic acid	HCN	—	5.6	40.0	−17.8	538
Esters						
Methyl formate	$C_2H_4O_2$	−1003.0	5.9	22.0	−19.0	236
Ethyl formate	$C_3H_6O_2$	−1638.8	2.7	16.4	−20.0	577
Methyl acetate	$C_3H_6O_2$	−1628.1	3.1	16.0	−9.4	654
Ethyl acetate	$C_4H_8O_2$	−2273.6	2.5	9.0	−4.4	610
Propyl acetate	$C_5H_{10}O_2$	—	2.0	8.0	14.4	662
Isopropyl acetate	$C_5H_{10}O_2$	−2907.0	1.8	8.0	—	572
Butyl acetate	$C_6H_{12}O_2$	−3587.8	1.7	7.6	22.2	423
Amyl acetate	$C_7H_{14}O_2$	−4361.7	1.1	—		399
Inorganic						
Hydrogen	H_2	−285.8	4.0	75.0	—	572
Ammonia	NH_3	−382.6	15.0	28.0	—	651
Cyanogen	C_2N_2	−1080.7	6.0	32.0	—	850
Oxides						
Ethylene oxide	C_2H_4O	−1264.0	3.0	80.0	−20.0	429
Propylene oxide	C_3H_6O	—	2.0	22.0	−37.2	748
Dioxan	$C_4H_8O_2$	—	2.0	22.0	12.2	266
Sulfides						
Carbon disulfide	CS_2	−1031.8	1.2	44.0	−30.0	149
Hydrogen sulfide	H_2S	−562.6	4.3	45.0	—	292
Carbon oxysulfide	COS	−546.0	12.0	29.0	—	—

Compound	Formula	Heat of Combustion kJ/mol	Flammability Limit Vol. % Fuel in air		Flashpoint Temp.°C	Autoignition Temp. °C
			LFL	UFL		
Chlorides						
Methyl chloride	CH_3Cl	−687.0	10.7	17.4	0.0	632
Ethyl chloride	C_2H_5Cl	−1325.0	3.8	14.8	−50.0	516
Propyl chloride	C_3H_7Cl	−2001.3	2.6	11.1	<−17.7	520
Butyl chloride	C_4H_9Cl	—	1.8	10.1	−12.0	460
Isobutyl chloride	C_4H_9Cl	—	2.0	8.8	—	—
Allyl chloride	C_3H_9Cl	—	3.3	11.1	−31.7	487
Amyl chloride	$C_5H_{11}Cl$	—	1.6	8.6	—	259
Vinyl chloride	C_2H_3Cl	—	4.0	22.0	−8.0	—
Ethylene dichloride	$C_2H_2Cl_2$	−1133.8	6.2	16.0	—	413
Propylene dichloride	$C_3H_6Cl_2$	—	3.4	14.5	−51.7	557
Bromides						
Methyl bromide	CH_3Br	−768.9	13.5	14.5	−20.0	537
Ethyl bromide	C_2H_5Br	−1424.6	6.7	11.3	—	588
Allyl bromide	C_3H_5Br	—	4.4	7.3	—	295
Amines						
Methyl amine	CH_5N	−1085.1	4.9	20.7	0.0	430
Ethyl amine	C_2H_7N	−1739.9	3.5	14.0	—	384
Dimethyl amine	C_2H_7N	−1768.9	2.8	14.4	—	402
Propyl amine	C_3H_9N	−2396.6	2.0	10.4	—	318
Diethyl amine	$C_4H_{11}N$	−3074.3	1.8	10.1	—	312
Trimethyl amine	C_3H_9N	−2443.0	2.0	11.6	—	—
Triethyl amine	$C_6H_{15}N$	−4134.5	1.2	8.0	—	—

Notes:

Normal alkanes are denoted without the "n-."

*Heat of combustion data from Suzuki (1994). Flammability limits from Lewis and Von Elbe (1987). Flashpoint data from Sax (1984). Autoignition data from Glassman (1996).

PROCEDURE FOR EXAMPLE 3.2

The following procedure is the solution to Example 3.2 found in Section 3.11. This procedure is formatted according to methods described in the AIChE CCPS book titled *Guidelines for Writing Effective Operating and Maintenance Procedures* (AIChE, 1996c). The apparatus in question is shown in Figure 3.16. The configuration for the vessel charging is shown in Figure 3.17 and the configuration for transferring the liquid product from the vessel to a drum is shown in Figure 3.18.

Please refer to Example 3.2 and the figures for additional supporting information.

The author thanks Bob Walter for developing this written procedure.

CCPS Chemicals Inc.	OPS-NO-001 Rev. 0
Batch Processing Division	Operating Procedure

CHARGING METHYL ETHYL KETONE (MEK) TO REACTOR (R-2)
AND DRUMMING FINAL PRODUCT

APPROVED BY: _____   DATE : _____

                              Manager

Table of Contents

PURPOSE

This procedure describes how to charge Methyl Ethyl Ketone (MEK) from a 55-gallon drum into Reactor (R-2) using vacuum. After additional custom batch instructions have been completed, this procedure describes how to transfer processed material from (R-2) into 55-gallon drums. (R-2) is a 100-gallon, jacketed, glass-lined reactor pressure rated at 25 psig at 650°F. Reactor safety features include:

- A sight glass

- 2-inch safety valve set at 25 psig

- 2-inch 316 stainless steel rupture disc set at 22 psig

CHARGING METHYL ETHYL KETONE (MEK) TO REACTOR (R-2) AND DRUMMING FINAL PRODUCT

REFERENCES

CCPS Chemical Co., Safe Work Practices Manual

- *PPE / Job Requirements Matrix*, SWP-003
- *Bonding and Grounding Guidelines*, SWP-014
- *Using the Plant Nitrogen System*, SWP-015

CCPS Chemical Co., Equipment Operating Procedures

- *Cleaning Reactor (R-2)*, OPS-NO-013

Process Safety Information Package

- *Reactor R-2 and Auxiliary Equipment*, PT-011

CCPS Electronic Material Safety Data Sheet Software

The CCPS computer network provides access to an electronic database of MSDSs for all chemicals approved for use on the site. Always refer to the MSDS for details about materials in the workplace. Performing this procedure may cause exposure to:

- Methyl Ethyl Ketone – MEK
- Nitrogen – N_2

Applicable Batch Instruction Procedures

- *As directed by the Operations Supervisor*

CHARGING METHYL ETHYL KETONE (MEK) TO REACTOR (R-2)
AND DRUMMING FINAL PRODUCT

PRECAUTIONS

Health Hazards   Methyl Ethyl Ketone (MEK) is a volatile solvent. MEK can affect you when
breathed in and by passing through your skin.

MEK should be handled as a teratogen--with extreme caution.

Exposure can cause dizziness, headache, blurred vision, and cause you to pass
out.

Repeated exposures, along with other solvents, can damage the nervous system.

The liquid can severely burn the eyes and may irritate the skin. Repeated
exposure can cause drying and cracking of the skin. The vapor can irritate the
eyes, nose, mouth, and throat.

Avoid breathing vapors, skin and eye contact by using caution, ventilation and
appropriate PPE.

Nitrogen is an asphyxiant. Avoid confined spaces where adequate ventilation is
not provided.

Flammability and   MEK has a flash point of 20°F. Vapors are flammable! Use inerting, grounding
Explosivity Hazards and bonding techniques per SWP-014, *Bonding and Grounding Guidelines*.
Summary MSDS Information on MEK:

Characteristic	Limit
Flashpoint:	20°F (-6.7°C)
LFL:	1.8%
UFL:	10.0%
Boiling point:	176°F (80°C)
TLV-TWA:	200 ppm

CCPS Chemicals Inc.   OPS-NO-001 Rev. 0
Batch Processing Division   Operating Procedure

CHARGING METHYL ETHYL KETONE (MEK) TO REACTOR (R-2)
AND DRUMMING FINAL PRODUCT

Precautions (cont.)

MEK is a stable material in closed containers at room temperature under normal storage and handling conditions.

MEK is a Class IB flammable liquid. It must only be transferred in a Class I, Division 1 electrical area. Dilution ventilation must be present at the rate of 1 ft^3/min per ft^2 of floor area.

Safety Systems   An open-head sprinkler system provides deluge to the process area and is
Information   activated by a flammable vapor detector set at 25% of the LFL.

The vessel will be pressurized with nitrogen then vacuum purged to meet a target concentration. The minimum oxygen concentration for MEK is 11% by volume oxygen. This is slightly below the in-service oxygen concentration of 11.5%.

SPECIAL TOOLS AND EQUIPMENT

- Hardhat
- Safety shoes
- Face shield
- MEK resistant apron or coat and gloves
- Non-sparking bung wrench
- Non-sparking T-fitting for MEK drum nitrogen pad
- Ventilation equipment operating at no less than 1 ft^3 per ft^2 of floor area
- Special ventilation (elephant trunk) near top of drum is required unless operation is occurring in the open.
- 1 bonding strap with type A bolt and type B spring loaded clamp at each end.
- 1 bonding strap with type B spring loaded clamps on each end.
- Conductive drumming hose with drum dip pipe
- Portable Oxygen Monitor

CCPS Chemicals Inc.   OPS-NO-001 Rev. 0
Batch Processing Division   Operating Procedure

CHARGING METHYL ETHYL KETONE (MEK) TO REACTOR (R-2)
AND DRUMMING FINAL PRODUCT

PREREQUISITES

1. Verify the Quality Control lab has confirmed that the MEK in the drum is as stated on the MSDS.

2. Verify ventilation system is operating.

3. Verify calibration check dates are within the last 12 months for:
   - 2-inch safety valve set at 25 psig
   - 2-inch 316 stainless steel rupture disc set at 22 psig

4. Verify reactor (R-2) has been hydrostatically tested to 25 psig within the last 12 months.

5. Verify process is grounded to building ground system and grounding was checked within the last 12 months.

6. Verify (R-2) is empty and washed.

7. Physically verify integrity of all bonding and grounding straps and clamps.

CHARGING METHYL ETHYL KETONE (MEK) TO REACTOR (R-2) AND DRUMMING FINAL PRODUCT

PROCEDURE

Preparing Reactor (R-2) for MEK Transfer

Ensure Initial   1. Verify the reactor manway is closed with all bolts tightened.
Configuration

2. Verify initial reactor valve configuration is as follows:

Valve Name and ID Number	Position
Auxiliary Purge/Vent Valve (V-1)	Closed
Dip Leg Valve (V-2)	Closed
Manifold Line Valve (V-3)	Closed
Manual Vent Valve (V-4)	Closed
Evacuation Valve (V-5)	Closed
Nitrogen Inlet Valve (V-6)	Closed
Auxiliary Purge/Vent Valve (V-7)	Closed
Pressure Gauge Isolation Valve (V-8)	Open
Purge Nitrogen Inlet Valve (V-9)	Closed
Purge Line Block Valve (V-10)	Closed
Drum Out Line Purge Valve (V-11)	Closed
Drum Out Line Block Valve (V-12)	Closed
Drum Out Valve (V-13)	Closed

3. Open evacuation valve (V-5) to begin evacuating reactor (R-2).

4. Begin monitoring pressure gauge (P-1).

5. When (P-1) reads 0.058 psia for 3 minutes, close evacuation valve (V-5).

6. Open nitrogen inlet valve (V-6).'

7. Begin monitoring pressure gauge (P-1).

Preparing Reactor (R-2) for MEK Transfer (cont.)

CHARGING METHYL ETHYL KETONE (MEK) TO REACTOR (R-2)
AND DRUMMING FINAL PRODUCT

8. When pressure gauge (P-1) reads above 30 psia for 3 minutes, close valve (V-6).

9. Open evacuation valve (V-5) to begin evacuating reactor (R-2).

10. Begin monitoring pressure gauge (P-1).

11. When (P-1) reads 0.058 psia for 3 minutes, close evacuation valve (V-5).

12. Monitor pressure gauge (P-1) after 5 minutes.

13. Has pressure gauge (P-1) reading risen above 0.058 psia?
    - **YES – STOP**. Check reactor (R-2) for leaks. Notify maintenance to check and repair all leaks.
    - **NO – Continue.**

[!]   **CAUTION**

Fire and explosion hazard. Ensure installation of conductive charging hose is made to dip leg valve (V-2) rather than (V-1) or (V-7). Incorrect installation could increase likelihood of static charge buildup due to liquid free fall.

---

14. Attach conductive charging hose to dip leg valve (V-2).

CCPS Chemicals Inc.   OPS-NO-001 Rev. 0
Batch Processing Division   Operating Procedure

CHARGING METHYL ETHYL KETONE (MEK) TO REACTOR (R-2)
AND DRUMMING FINAL PRODUCT

Preparing the MEK Drum for Transfer

1. Place drum in position for the transfer.

2. Bond the MEK drum to the reactor:
   - Use a type A bolt clamp at the reactor
   - Use a type B spring clamp at the drum

3. Place elephant trunk ventilation hose into position near top of the MEK drum.

4. Open both bungs on the MEK drum using non-sparking bung wrench.

5. Install T-fitting on òne bunghole to enable nitrogen padding the MEK drum.

6. Open the nitrogen valve to the T-fitting to begin nitrogen pad on the MEK drum.

7. Insert the flexible hose dip pipe connected to dip leg valve (V-2) into the remaining MEK drum bunghole.

8. Verify the tip of the dip pipe is at the bottom of the MEK drum.

9. Bond the dip pipe to the drum using type B spring loaded clamps on each end of the bonding strap.

10. Open the valve on the flexible hose MEK drum dip pipe.

CCPS Chemicals Inc.   OPS-NO-001 Rev. 0
Batch Processing Division   Operating Procedure

**CHARGING METHYL ETHYL KETONE (MEK) TO REACTOR (R-2)
AND DRUMMING FINAL PRODUCT**

Charging MEK from the Drum to the Reactor

☐❗

CAUTION

Potential exposure to MEK during the next steps. Wear appropriate PPE and avoid contact with MEK vapors and liquid.

CAUTION

Fire and explosion hazard. An explosive atmosphere is possible in the reactor if the MEK drum dip pipe valve is not closed immediately after all MEK is emptied. Air might enter the reactor creating an explosive atmosphere.

1. Open reactor dip leg valve (V-2) to begin MEK flow from the drum to reactor (R-2).

2. Throttle the valve on the flexible hose MEK drum dip pipe to adjust MEK flow.

3. Monitor MEK transfer.

4. When MEK charging is complete, lift the flexible hose MEK drum dip pipe to allow liquid in the transfer line to be drawn into the reactor.

5. Close the flexible hose MEK drum dip pipe valve.

6. Close the reactor dip leg valve (V-2).

7. Open nitrogen valve (V-6) to pressurize reactor (R-2).

CHARGING METHYL ETHYL KETONE (MEK) TO REACTOR (R-2)
AND DRUMMING FINAL PRODUCT

Removing and Cleaning MEK Drum

1. Disconnect the spring-loaded type B bonding clamp from the flexible hose MEK drum dip pipe.

2. Remove dip pipe from used MEK drum and place in drip collection tube.

3. Close the nitrogen valve to the MEK drum T fitting to stop nitrogen padding to the drum.

4. Remove the T fitting from the MEK drum bunghole.

5. Replace both MEK drum bungs and verify they are tightly sealed.

6. Remove the spring-loaded type B clamp end of the bonding strap from the drum.

7. Move the used MEK drum to the cleaning and disposal area.

8. Notify the Operations Supervisor transfer is complete and request the batch instructions assigned to reactor (R-2).

9. Review batch instructions and perform the required operations.

**CHARGING METHYL ETHYL KETONE (MEK) TO REACTOR (R-2)
AND DRUMMING FINAL PRODUCT**

Transfer Product from Reactor (R-2) to a Drum

Preparing the   1.   At the workstation beneath reactor (R-2), connect a conductive drumming
     Reactor   hose with a drum dip pipe to drum out valve (V-13).

Preparing the   2.   Place a clean product drum into position at the workstation.
receiving Drum

             3.   Connect a bonding strap to the reactor using a type A bolt clamp at the
                    reactor end and a type B spring-loaded clamp at the drum end.

             4.   Using a non-sparking bung wrench, open both bungs on the drum.

             5.   Insert the flexible nitrogen hose into one bunghole, ensuring it rests on the
                    bottom of the drum.

             6.   Adjust nitrogen flow into the drum at the workstation regulator to $3 - 5$ ft^3
                    per minute.

             7.   After 10 minutes, measure oxygen concentration at the flow coming from
                    the second bunghole.

             8.   Repeat measurements regularly until oxygen level is undetectable. The drum
                    is now inerted.

             9.   Close the regulator valve to stop nitrogen flow to the drum and remove the
                    flexible nitrogen hose.

            10. Insert a drum dip pipe into a bunghole ensuring it reaches to the bottom of
                    the drum.

CHARGING METHYL ETHYL KETONE (MEK) TO REACTOR (R-2)
AND DRUMMING FINAL PRODUCT

Transfer Product from Reactor (R-2) to a Drum (cont.)

11. Connect a bond strap from the dip pipe to the drum using a bonding strap with a type B spring loaded clamps at each end.

CAUTION

Potential exposure to MEK during the next steps. Wear appropriate PPE and avoid contact with MEK vapors and liquid.

CAUTION

Fire and explosion hazard. Air may re-enter the drum and create a flammable atmosphere if significant time elapses between the end of the nitrogen purge and the filling of the drum. Additional purging is required if this occurs.

Transferring the Material

12. Open the drum out line block valve (V-12).

13. Open the drum out valve (V-13) to begin draining material into the drum.

14. Monitor level in the drum visually through the open bunghole.

15. When drum is filled to intended capacity, ensure reactor (R-2) contents are completely drained into the drum or into the work station waste line.

16. Close drum out line block valve (V-12).

17. Close drum out valve (V-13).

Disconnecting the Drum

18. Disconnect the bonding strap from the drum dip pipe.

19. Remove the dip pipe from the bunghole and place in drip collection tube.

20. Using a nonsparking bung wrench, replace both bungs in the drum.

21. Disconnect the bonding strap from the drum.

**CHARGING METHYL ETHYL KETONE (MEK) TO REACTOR (R-2)
AND DRUMMING FINAL PRODUCT**

Transfer Product from Reactor (R-2) to a Drum (cont.)

22. Apply the label provided with the batch instruction to the drum.

23. Move drum to the storage and shipping area.

24. If this transfer completes the campaign for the product, perform procedure *Cleaning Reactor (R-2)*, OPS-NO-013.

END

Appendix E

COMBUSTION DATA FOR DUST CLOUDS[a]

Dust	Median Particle Size (mm)	Minimum Explosive Dust Conc. (gm/m³)	P_{max} (barg)	K_{St} (bar-m/sec)	Minimum Ignition Energy (mJ)
Cotton, Wood, Peat					
Cotton	44	100	7.2	24	—[b]
Cellulose	51	60	9.3	66	250
Wood dust	33	—	—	—	100
Wood dust	80	—	—	—	7
Paper dust	<10	—	5.7	18	—
Feed, Food					
Dextrose	80	60	4.3	18	—
Fructose	200	125	6.4	27	180
Fructose	400	—	—	—	>4000
Wheat grain dust	80	60	9.3	112	—
Milk powder	165	60	8.1	90	75
Rice flour	—	60	7.4	57	>100
Wheat flour	50	—	—	—	540
Milk sugar	10	60	8.3	75	14
Coal, Coal Products					
Activated Carbon	18	60	8.8	44	—
Bituminous coal	<10	—	9.0	55	—
Plastics, Resins, Rubber					
Polyacrylamide	10	250	5.9	12	—
Polyester	<10	—	10.1	194	—
Polyethylene	72	—	7.5	67	—
Polyethylene	280	—	6.2	20	—

Dust	Median Particle Size (mm)	Minimum Explosive Dust Conc. (gm/m^3)	P_{max} (barg)	K_{St} (bar-m/sec)	Minimum Ignition Energy (mJ)
Polypropylene	25	30	8.4	101	—
Polypropylene	162	200	7.7	38	—
Polystyrene (copolymer)	155	30	8.4	110	—
Polystyrene (hard foam)	760	—	8.4	23	—
Polyurethane	3	<30	7.8	156	—
Intermediate products, Auxiliary materials					
Adipinic acid	<10	60	8.0	97	—
Naphthalene	95	15	8.5	178	<1
Salicylic acid	—	30	—	—	—
Other technical, Chemical products					
Organic dyestuff (blue)	<10	—	9.0	73	—
Organic dyestuff (red)	<10	50	11.2	249	—
Organic dyestuff (red)	52	60	9.8	237	—
Metals, Alloys					
Aluminum powder	<10	60	11.2	515	—
Aluminum powder	22	30	11.5	110	—
Bronze powder	18	750	4.1	31	—
Iron (from dry filter)	12	500	5.2	50	—
Magnesium	28	30	17.5	508	—
Magnesium	240	500	7.0	12	—
Zinc (dust from collector)	<10	250	6.7	125	—
Other inorganic products					
Graphite (99.5% C)	7	<30	5.9	71	—
Sulfur	20	30	6.8	151	—
Toner	<10	60	8.9	196	4

[a]From Eckhoff (1997)

[b]The symbol "—" means that data are not available.

REFERENCES

ACGIH (1998), *Industrial Ventilation, A Manual of Recommended Practice*, 23rd ed, Cincinnati, OH, American Conference of Governmental Industrial Hygienists.

AIChE (1989), *Guidelines for Chemical Process Quantitative Risk Analysis*, New York: American Institute of Chemical Engineers.

AIChE (1992a), *Guidelines for Hazard Evaluation Procedures*, New York: American Institute of Chemical Engineers.

AIChE (1992b), *Guidelines for Investigating Chemical Process Incidents*, New York: American Institute of Chemical Engineers.

AIChE (1993), *Guidelines for Engineering Design for Process Safety*, New York: American Institute of Chemical Engineers.

AIChE (1994), *Guidelines for Evaluating the Characteristics of Vapor Cloud Explosions, Flash Fires, and BLEVEs*, New York: American Institute of Chemical Engineers.

AIChE (1995a), *Guidelines for Chemical Reactivity Evaluation and Application to Process Design*, New York: American Institute of Chemical Engineers.

AIChE (1995b), *International Symposium on Runaway Reactions and Pressure Relief Design*, Boston, MA. New York: American Institute of Chemical Engineers.

AIChE (1995c), *Guidelines for Process Safety Fundamentals in General Plant Operations*, New York: American Institute of Chemical Engineers.

AIChE (1996a), *Guidelines for Use of Vapor Cloud Dispersion Models*, 2nd ed, New York: American Institute of Chemical Engineers.

AIChE (1996b), *Guidelines for Evaluating Process Plant Buildings for External Explosions and Fires*, New York: American Institute of Chemical Engineers.

AIChE (1996c), *Guidelines for Writing Effective Operating and Maintenance Procedures*, New York: American Institute of Chemical Engineers.

AIChE (1997), *Guidelines for Postrelease Mitigation Technology in the Chemical Process Industry*, New York: American Institute of Chemical Engineers.

AIChE (1998), *Guidelines for Pressure Relief and Effluent Systems*, New York: American Institute of Chemical Engineers.

AIChE (1999a), *Guidelines for Consequence Analysis of Chemical Releases*, New York: American Institute of Chemical Engineers.

AIChE (1999b), *Estimating the Flammable Mass of a Vapor Cloud*, New York: American Institute of Chemical Engineers.

AIChE (2000), *Guidelines for Chemical Process Quantitative Risk Analysis*, 2nd ed, New York: American Institute of Chemical Engineers.

API 500 (1998), *Classification of Locations for Electrical Installation*, Washington, DC: American Petroleum Institute.

API 520 (2000), *Design, Installation of Pressure Relieving Systems*, Washington, DC: American Petroleum Institute.

API 521 (1999), *Guide for Pressure-Relieving and Depressuring Systems*, Washington, DC: American Petroleum Institute.

API 540 (1999), *Electrical Installations in Petroleum Processing Plants*, Washington, DC: American Petroleum Institute.

API 752 (1995), *Management of Hazards Associated with Location of Process Plant Buildings*, Washington, DC: American Petroleum Institute.

API 2000 (1998), *API 2000: Venting Atmospheric and Low Pressure Storage Tanks: Nonrefrigerated and Regrigerated*, 5th ed, Washington, DC: American Petroleum Institute.

API 2003 (1998), *Protection Against Ignitions Arising Out of Static, Lightning and Stray Currents*, Washington, DC: American Petroleum Institute.

API 2030 (1998), *Application of Water Spray Systems for Fire Protection in the Petroleum Industry*, Washington, DC: American Petroleum Institute.

API 2218 (1999), *Fireproofing Practices in Petroleum and Petrochemical Processing Plants*, Washington, DC: American Petroleum Institute.

ASCE (1997), *Design of Blast Resistant Buildings in Petrochemical Facilities*, Reston, VA: American Society of Civil Engineers.

ASTM D3278 (1989), *ASTM D3278: Standard Test Methods for flashpoint of Liquids by Setaflash Closed-Cup Apparatus*, Philadelphia: American Society for Testing and Materials.

ASTM D5687 (1987), *Standard Test Method for flashpoint by Tag Closed Tester*, Philadelphia, PA: American Institute for Testing and Materials.

ASTM D9290 (1990a), *ASTM D9290: Standard Test Method for Flash and Fire Points by Cleveland Open Cup*, Philadelphia: American Society for Testing and Materials.

ASTM D9390 (1990b), *ASTM D9390: Standard Test Methods for flashpoint by Pensky-Martens Closed Tester*, Philadelphia: American Society for Testing and Materials.

ASTM E918 (1992), *ASTM E918: Standard Practice for Determining Flammability Limits at Elevated Temperature and Pressure*, Philadelphia: American Society for Testing and Materials.

Baker, Q. A. (2003), *Personal communication*.

Baker, Q. A., C. M. Doolittle, et al. (1997), *Recent Developments in the Baker-Strehlow VCE Analysis Methodology*, 31st Loss Prevention Symposium, Houston, TX. New York: American Institute of Chemical Engineers.

Baker, Q. A., M. J. Tang, et al. (1994), *Vapor Cloud Explosion Analysis*, 28th Loss Prevention Symposium, Atlanta, GA. New York: American Institute of Chemical Engineers.

Baker, W. E. (1983), *Explosions in Air*, San Antonio: Baker Engineering and Risk Consultants.

Baker, W. E., P. A. Cox, et al. (1988), *Explosion Hazards and Evaluation*, New York: Elsevier Scientific Publishing Co.

Bartknecht, W. (1981), *Explosions: Course, Prevention, Protection*, Berlin: Springer-Verlag.

Barknecht, W. (1993), *Explosionschutz*, New York: Springer-Verlag.

Benuzzi, A. and J. M. Zaldivar, Eds. (1991). *Safety of Chemical Batch Reactors and Storage Tanks*, Dordrecht, The Netherlands: Kluwer Academic Publishers.

Bollinger, R. E., D. G. Clark, et al. (1996), *Inherently Safer Chemical Processes: A Life Cycle Approach*, New York: American Institute of Chemical Engineers.

Bond, J. (1991), *Sources of Ignition*, Oxford: Butterworth-Heinemann.

Brasie, W. C. and D. W. Simpson (1968), *Guidelines for Estimating Damage Explosion*, 2nd Loss Prevention Symposium, New York: American Institute of Chemical Engineers.

Britton, L. G. (1993), "Static Hazards Using Flexible Intermediate Bulk Containers for Powder Handling" *Process Safety Progress*, **12**(4): 240–250.

Britton, L. G. (1999), *Avoiding Static Ignition Hazards in Chemical Operations*, New York: American Institute of Chemical Engineers.

Britton, L. G. (2002a), "Two Hundred Years of Flammability Limits" *Process Safety Progress*, **21**(1).

Britton, L. G. (2002b), "Using Heats of Oxidation to Evaluate Flammability Hazards" *Process Safety Progress*, **21**(1).

Brode, H. L. (1959), "Blast Waves from a Spherical Charge" *Phys. Fluids*, **2**: 217.

Brown, S. J. (1985), "Energy Release Protection for Pressurized Systems. Part I: Review of Studies into Blast and Fragmentation" *Appl Mech Rev*, **38**(12).

Brown, S. J. (1986), "Energy Release Protection for Pressurized Systems. Part II: Review of Studies into Impact/Terminal Ballistics" *Appl Mech Rev*, **39**(2).

Buettner, K. J. K. (1957), *Paper 57-SA-20: Heat Transfer and Safe Exposure Time for Man in Extreme Thermal Environments*, New York: American Society of Mechanical Engineers.

Checkel, M. D., D. S. K. Ting, et al. (1995), "Flammability Limits and Burning Velocities of Ammonia/nitric Oxide Mixtures" *J. Loss Prev. Process Ind.*, **8**(4): 215–220.

Clancey, V. J. (1972), *Diagnostic Features of Explosion Damage*, 6th International Meeting on Forensic Sciences, Edinburgh, Scotland.

Clayton, W. E. and M. L. Griffin (1994), *Catastrophic Failure of a Liquid Carbon Dioxide Storage Vessel*, 28th Loss Prevention Symposium, Atlanta, GA. New York: American Institute of Chemical Engineers.

Crouse Hinds (1999), *1999 Code Digest*, Syracuse, NY: Crouse Hinds Electrical Equipment.

Crowl, D. A. (1992), "Calculating the Energy of Explosion Using Thermodynamic Availability" *J. Loss prev. Process Ind.*, **5**(2): 109–118.

Crowl, D. A. and J. F. Louvar (1990), *Chemical Process Safety: Fundamentals with Applications*, Englewood Cliffs, NJ: Prentice Hall.

Crowl, D. A. and J. F. Louvar (2002), *Chemical Process Safety: Fundamentals with Applications*, 2nd ed, Upper Saddle River, NJ: Prentice Hall.

Decker, D. A. (1974), *An Analytical Method for Estimating Overpressure from Theoretical Atmospheric Explosions*, NFPA Annual Meeting, Quincy, MA: National Fire Protection Association.

DOT (2000), *49CFR173.120: Transportation*, Washington, DC: U. S. Department of Transportation.

Eckhoff, R. K. (1997), *Dust Explosions in the Process Industries*, Oxford: Butterworth-Heinemann.

Eggen, J. B. M. M. (1998), *HSE Report 202: GAME—Development of Guidance for the Application of the Multi-Energy Method*, London: Health and Safety Executive.

Eichel, F. G. (1967), "Electrostatics" *Chemical Engineering*: March 13, p. 153.

Eisenberg, N. A., C. J. Lynch, et al. (1975), *CG-D-136-75 and NTIS AD-015-245: Vulnerability Models: A Simulation System for Assessing Damage Resulting from Marine Spills*, Springfield, VA: U. S. Coast Guard Office of Research and Development.

ESCIS (1993), *Thermal Process Safety*, Basel, Switzerland: Expert Commission for Safety in the Swiss Chemical Industry.

Expert Commission for Safety in the Swiss Chemical Industry (1988), "Static Electricity: Rules for Plant Safety" *Plant/Operations Progress*: January, p. 19.

Fisher, H. G., H. S. Forrest, et al. (1992), *Emergency Relief System Design Using DIERS Technology*, New York: American Institute of Chemical Engineers.

Fthenakis, V. M., Ed. (1993). *Prevention and Control of Accidental Releases of Hazardous Gases*, New York: Van Nostrand.

Glassman, I. (1996), *Combustion*, 3rd ed., San Diego: Academic Press.

Glasstone, S., Ed. (1962). *The Effects of Nuclear Weapons*, Washington, DC: United States Atomic Energy Commission.

Grossel, S. S. (2002), *Deflagration and Detonation Flame Arresters*, New York: American Institute of Chemical Engineers.

Gugan, K. (1979), *Unconfined Vapour Cloud Explosions*, Houston: Gulf Publishing.

Hanna, S. R. and R. E. Britter (2002), *Wind Flow and Vapor Cloud Dispersion at Industrial and Urban Sites*, New York: American Institute of Chemical Engineers.

Hansel, J. G., J. W. Mitchell, et al. (1991), *Predicting and Controlling Flammability of Multiple Fuel and Multiple Inert Mixtures*, AIChE Loss Prevention Symposium, Pittsburgh. New York: American Institute of Chemical Engineers.

Henley, B. F. (1998), "A Model for the Calculation and the Verification of Closed Cup flashpoints for Multi-component Mixtures" *Process Safety Progress*, 17: 86–87.

Hougen, O. A., K. M. Watson, et al. (1954), *Chemical Process Principles, Part 1: Material and Energy Balances*, 2nd ed, New York: John Wiley and Sons, Inc.

Hymes, J. (1983), *Report SRD R 275:The Physiological and Pathological Effects of Thermal Radiation*, U.K. Atomic Energy Authority.

Jones, G. W. (1938), "Inflammation Limits and Their Practical Application in Hazardous Industrial Operations" *Chem. Rev.*, 22(1): 1–26.

Jones, T. B. and J. L. King (1991), *Powder Handling and Electrostatics*, Chelsea, MI: Lewis Publishers.

Kinney, G. F. and K. J. Graham (1985), *Explosive Shocks in Air*, Berlin: Springer-Verlag.

Kinsella, K. G. (1993), *A Rapid Assessment Technology for the Prediction of Vapor Cloud Explosion Overpressure*, International Conference and Exhibition on the Safety, Health and Loss Prevention in the Oil, Chemical and Process Industries, Singapore.

Kletz, T. (1991), *Plant Design for Safety: A User-Friendly Approach*, New York: Hemisphere Publishing.

Larson, E. D. (1998), *Flashpoint Depression in Minimum Boiling Point Azeotropic Mixtures*, Houghton, MI: MS Thesis, Michigan Technological University.

Le Chatelier, H. (1891), "Estimation of Firedamp by Flammability Limits" *Ann. Mines,* **19**(8): 388–395.

Lees, F. P. (1986), *Loss Prevention in the Process Industries*, London, Butterworths.

Lees, F. P. (1996), *Loss Prevention in the Process Industries*, 2nd ed., Oxford: Butterworth Heinemann.

Lenoir, E. M. P. and J. A. Davenport (1992), *A Survey of Vapor Cloud Explosions—Second Update*, 26th Loss Prevention Symposium, New Orleans, LA. New York: American Insitute of Chemical Engineers.

Leung, J. (1997), *Lecture on Reactivity*, SACHE Faculty Workshop, Wyandotte, MI. New York: American Institute of Chemical Engineers.

Lewis, B. and G. von Elbe (1987), *Combustion, Flames and Explosions of Gases*, Orlando: Academic Press.

Louvar, J. F., B. Maurer, et al. (1994), "Tame Static Electricity" *Chemical Engineering Progress,* **90**(11): 75.

M&M Protection Consultants (1998), *Large Property Damage Losses in the Hydrocarbon—Chemical Industries: A Thirty Year Review*, 15th ed, Chicago: Marsh and McLennan.

Mashuga, C. V. and D. A. Crowl (1998), "Application of the Flammability Diagram for Evaluation of Fire and Explosion Hazards of Flammable Vapors," 32nd Loss Prevention Symposium, New Orleans. New York: American Institute of Chemical Engineers.

Mashuga, C. V. and D. A. Crowl (1999), "Flammability Zone Prediction Using Calculated Adiabatic Flame Temperatures," 33rd Loss Prevention Symposium, Houston, TX. New York: American Institute of Chemical Engineers.

Melhem, G. A. (1997), "A Detailed Method for Estimating Mixture Flammability Limits Using Chemical Equilibrium" *Process Safety Progress,* **16**(4): 203–218.

Mercx, W. P. M., A. C. van den Berg, et al. (1998), *TNO Report PML 1998-C53: Application of Correlations to Quantify the Source Strength of Vapor Cloud Explosions in Realistic Situations*, The Netherlands: TNO Prins Maurits Laboratory.

Mixter, G. (1954), *Report UR-316: The Empirical Relation Between Time and Intensity of Applied Thermal Energy in Production of 2+ Burns in Pigs*, Rochester, NY: University of Rochester.

Mudan, K. S. (1984), "Thermal Radiation Hazards from Hydrocarbon Pool Fires" *Proc Energy Combust Sci,* **10**(1): 59–80.

Mudan, K. S. and P. A. Croce (1988), *Fire Hazard Calculations for Large Open Hydrocarbon Fires*, SFPE Handbook of Fire Protection Engineering, Boston, MA: Society of Fire Protection Engineers.

National Safety Council (1974), *Accident Prevention Manual for Industrial Operations*, Chicago: National Safety Council.

NFPA 13 (1999), *Standard for the Installation of Sprinkler Systems*, Quincy, MA: National Fire Protection Association.

NFPA 15 (2001), *Standard for Water Spray Fixed Systems for Fire Protection*, Quincy, MA: National Fire Protection Association.

NFPA 25 (2002), *Standard for the Inspection, Testing, and Maintenance of Water-Based Fire Protection Systems*, Quincy, MA: National Fire Protection Association.

NFPA 30 (2002), *Flammable and Combustible Liquids Code*, Quincy, MA: National Fire Protection Association.

NFPA 33 (2000), *Standard for Spray Application Using Flammable or Combustible Materials*, Quincy, MA: National Fire Protection Association.

NFPA 34 (2000), *Standard for Dipping and Coating Processes Using Flammable and Combustible Liquids*, Quincy, MA: National Fire Protection Association.

NFPA 58 (2001), *Liquefied Petroleum Gas Code*, Quincy, MA: National Fire Protection Association.

NFPA 59 (2001), *Utility LP-Gas Plant Code*, Quincy, MA: National Fire Protection Association.

NFPA 61 (1999), *Standard for the Prevention of Fires and Dust Explosions in Agricultural and Food Products Facilities*, Qunicy, MA, National Fire Protection Association.

NFPA 68 (1994), *NFPA 68: Venting of Deflagrations*, Qunicy, MA, National Fire Protection Association.

NFPA 68 (2002), *NFPA 68: Guide for Venting of Deflagrations*, Quincy, MA: National Fire Protection Association.

NFPA 69 (1997), *NFPA 69: Standard on Explosion Prevention Systems*, Quincy, MA: National Fire Protection Association.

NFPA 70 (2002), *National Electrical Code*, Quincy, MA: National Fire Protection Association.

NFPA 77 (2000), *NFPA 77: Recommended Practice on Static Electricity*, Quincy, MA: National Fire Protection Association.

NFPA 86 (1999), *Standard for Ovens and Furnaces*, Quincy, MA: National Fire Protection Association.

NFPA 120 (1999), *Standard for Coal Preparation Plants*, Qunicy, MA, National Fire Protection Association.

NFPA 230 (1999), *Standard for Fire Protection of Storage*, Qunicy, MA: National Fire Protection Association.

NFPA 480 (1998), *Standard for the Storage, Handling, and Processing of Magnesium Solids and Powders*, Quincy, MA: National Fire Protection Association.

NFPA 481 (2000), *Standard for the Production, Processing, Handling, and Storage of Titanium*, Quincy, MA: National Fire Protection Association.

NFPA 482 (1996), *Standard for the Production, Processing, Handling, and Storage of Zirconium*, Quincy, MA: National Fire Protection Association.

NFPA 496 (1998), *Standard for Purged and Pressurized Enclosures for Electrical Systems*, Quincy, MA: National Fire Protection Association.

NFPA 497 (1997), *Recommended Practice for the Classification of Flammable Liquids, Gases, or Vapors and of Hazardous (Classified) Locations for Electrical Installations in Chemical Process Areas*, Quincy, MA: National Fire Protection Association.

NFPA 499 (1997), *Recommended Practice for the Classification of Combustible Dusts and of Hazardous (Classified) Locations for Electrical Installations in Chemical Process Plants*, Quincy, MA: National Fire Protection Association.

NFPA 651 (1998), *Standard for the Machining and Finishing of Aluminum and the Production and Handling of Aluminum Powders*, Quincy, MA: National Fire Protection Association.

NFPA 654 (2000), *Standard for the Prevention of Fire and Dust Explosions from the Manufacturing, Processing and Handling of Combustible Particulate Solids*, Qunicy, MA, National Fire Protection Association.

NFPA 655 (2001), *Standard for Prevention of Sulfur Fires and Explosions*, Quincy, MA: National Fire Protection Association.

NFPA 664 (1998), *Standard for the Prevention of Fires and Explosions in Wood Processing and Woodworking Facilities*, Quincy, MA: National Fire Protection Association.

NFPA 780 (1997), *NFPA 780: Standard for the Installation of Lightning Protection Systems*, Quincy, MA: National Fire Protection Association.

NFPA (1999), *Automatic Sprinkler Systems Handbook*, Quincy, MA: National Fire Protection Association.

Olishifski, J. B. (1971), *Fundamentals of Industrial Hygiene*, 2nd ed, Chicago, IL, National Safety Council.

OSHA (1996), *OSHA 1910.106: Flammable and Combustible Liquids*, Washington, DC: Occupational Safety and Health Administration.

OSHA (1998), Web site: OSHA.gov, July, 1998, Occupational Safety and Health Administration, Washington, DC.

Oswald, C. J. (2001), *Report A150-301: Enhancement of the BEAST (Building Evaluation and Siting Tool) Computer Program*, San Antonio, TX, Baker Engineering and Risk Consultants.

Oswald, C. J. and Q. A. Baker (1999), "Vulnerability Model for Occupants of Blast Damaged Buildings," 34th Loss Prevention Symposium. New York: American Institute of Chemical Engineers.

Parker, R. J. (1975), *The Flixborough Disaster. Report of the Court of Inquiry*, London: HM Stationary Office.

Pratt, T. H. (1997), *Electrostatic Ignitions of Fires and Explosions*, New York: American Institute of Chemical Engineers.

Regenass, W. (1984), *The Control of Exothermic Reactions*, I.Chem.E Symposium Series, London: The Institution of Chemical Engineers.

Richard, S. (1986), "Inert Gas Blanketing for Safety and Moisture Control," ILTA National Operating Conference, Houston, TX.

Robinson, C. S. (1944), *Explosions, their Anatomy and Destructiveness*, New York: McGraw-Hill.

Sanders, R. E. (1999), *Chemical Process Safety: Learning from Case Histories*, Boston: Butterworth Heinemann.

Satyanarayana, K. and P. G. Rao (1992), "Improved Equation to Estimate flashpoints" *J. Haz. Mat.,* **32**: 81–85.

Sax, N. I. (1984), *Dangerous Properties of Industrial Materials*, 6th ed, New York: Van Nostrand.

Schubach, S. (1995), "Thermal Radiation Targets Use in Risk Analysis" *Trans IChemE,* **73**(Part B).

Senecal, J. A. and P. A. Beaulieu (1998), "K_G: New Data and Analysis" *Process Safety Progress,* **17**(1).

Smith, J. M. and H. C. Van Ness (1987), *Introduction to Chemical Engineering Thermodynamics*, 4th ed. New York: McGraw-Hill.

Stephens, M. M. (1970), *Minimizing Damage to Refineries*, Washington, DC, U. S. Department of Interior, Office of Oil and Gas.

Stoll, A. M. and L. C. Green (1958), "Paper 58-A-219: The Production of Burns by Thermal Radiation of Medium Intensity," New York: American Society of Mechanical Engineers.

Strehlow, R. A., R. T. Luckritz, et al. (1979), "The Blast Wave Generated by Spherical Flames" *Combustion and Flame,* **35**: 297-310.

Stull, D. R. (1977), *Fundamentals of Fires and Explosions*, AIChE Monograph Series 73 (No. 10), New York: American Institute of Chemical Engineers.

Suzuki, T. (1994), "Empirical Relationship between Lower Flammability Limits and Standard Enthalpies of Combustion of Organic Compounds" *Fire and Materials,* **18**: 333–336.

Suzuki, T. and K. Koide (1994), "Correlation between Upper Flammability Limits and Thermochemical Properties of Organic Compounds" *Fire and Materials,* **18**: 393–397.

Tang, M. J. and Q. A. Baker (1999), "A New Set of Blast Curves from Vapor Cloud Explosions" *Process Safety Progress,* **18**(4).

TM5-1300 (1990), *Structures to Resist the Effects of Accidental Explosions*, Washington, DC, Department of Army, Navy and Air Force.

TNO (1997), *TNO Yellow Book*, 3rd ed., Apeldoorn, The Netherlands: TNO.

USBL (1999), *USBL 99-208: National Census of Fatal Occupational Injuries, 1998*, Washington, DC: United States Department of Labor, Bureau of Labor Statistics.

USCSHIB (1998), *Investigation Report 98-007-IA: Propane Tank Explosion*, Washington, DC: U. S. Chemical Safety and Hazard Investigation Board.

van Winderden, C. J. M., A. C. van den Berg, et al. (1989), "Vapor Cloud Explosion Blast Prediction" *Plant/Operations Progress,* **8**(4): 234–238.

World Bank (1985), *Manual of Industrial Hazard Assessment Techniques*, Washington, DC: Office of Environmental and Scientific Affairs, World Bank.

Zabetakis, M. G. (1965), *Fire and Explosion Hazards at Temperature and Pressure Extremes*, AIChE Chem. Engr. Symp. Series 2, New York: American Institute of Chemical Engineers.

Zabetakis, M. G., S. Lambiris, et al. (1959), *Flame Temperatures of Limit Mixtures*, Seventh Symposium on Combustion, Boston: Butterworths.

Zeeuwen, J. P., and B. J. Wiekema (1978), "The Measurement of Relative Reactivities of Combustible Gases," Conference on Mechanisms of Explosions in Dispersed Energetic Materials.

GLOSSARY

Accident The occurrence of a sequence of events that produce unintended injury, death or property damage.

Adiabatic Flame Temperature The final temperature as a result of combustion without any heat losses.

Aerosol Liquid droplets or solid particles of size small enough to remain suspended in air for prolonged periods of time.

Arrival Time The time for a shock front to arrive at a fixed distance from an explosion.

Autoignition Temperature (AIT) A temperature above which a mixture will ignite without the need for an external ignition source—the mixture will appear to ignite spontaneously.

Blanketing An inerting method used to provide a continuously maintained atmosphere that is either inert or fuel rich in the vapor space of a container or vessel.

Blast Wave A pressure wave propagating in air as a result of an explosion.

BLEVE Boiling Liquid Expanding Vapor Explosion. An explosion that occurs when a vessel containing liquefied gas stored at a temperature above its normal boiling point fails catastrophically.

Bonding A procedure of connecting two process units together electrically in order to maintain the same electrical potential between the two units.

Brush Discharge A static electricity discharge occurring between a conductor and a nonconductor.

Bulking Brush Discharge *See* Conical Pile Discharge.

Burning Velocity The speed at which the flame front propagates relative to the unburned gas.

Charge Relaxation Time The time required for a static electricity charge to dissipate from an object.

Chemical Explosion An explosion due to a chemical reaction, including a combustion reaction, a decomposition reaction, or some other rapid exothermic reaction.

Combined Pressure–Vacuum Purging An inerting method that uses a vacuum and a pressurized source of inert gas to reduce the vapor concentration in a process to one that is not combustible.

Combustible A definition used by NFPA 30 (2000) for a liquid having a flash point temperature at or above 100°F.

Condensed Phase Explosion An explosion that occurs in the solid or liquid phase.

Conical Pile Discharge A static electricity discharge that occurs at the surface of a pile of powder as powder is being poured into a vessel.

Continuous Sweep An inerting method that uses a continuous flow of inert gas to reduce the oxygen concentration in a vessel such that the vapors are not combustible.

Corona Discharge A special case of a brush discharge, occurring between a charged nonconductor and a conductor with a small radius of curvature.

DDT *See* Deflagration to Detonation Transition.

Deflagration A reaction in which the speed of the reaction front propagates through the unreacted mass at a speed less than the speed of sound in the unreacted medium.

Deflagration to Detonation Transition (DDT) A deflagration that changes into a detonation.

Deluge System An open-head sprinkler system designed to discharge large quantities of water on to the surface of a process unit.

Detonable Limit A limiting fuel concentration at which a detonation can occur in a gas mixture.

Detonation A reaction in which the speed of the reaction front propagates through the unreacted medium at a speed greater than the speed of sound in the unreacted medium.

Dilution Ventilation A method of adding fresh air into a work area to reduce concentrations of flammable or toxic materials.

Dust Any finely divided solid, 420 μm or 0.016 inch, or less, in diameter (NFPA 68, 1998).

Dynamic Pressure A pressure caused by the wind associated with a blast wave.

Explosion A release of energy that causes a blast.

Explosion Proof Enclosures A specially designed housing containing an electrical fixture with arcing contacts, such as a switch, relay or contactor.

Fire A slow combustion that occurs without significant overpressures. Damage is mostly due to thermal and radiant energy release.

Fireball Burning of a large fuel–air cloud.

Fire Point The temperature above which a liquid is capable of producing enough vapor to form a flammable mixture that is capable of continuous combustion.

Flammable A term applied by NFPA 30 (2000) to liquids with a flash point below 100°F.

Flame Speed The speed at which the combustion wave moves relative to the unburned gas in the direction normal to the wave surface.

Flash Fire A combustion of a large cloud of flammable gas. No significant overpressures are produced, but damage may result from thermal radiation and direct flame impingement.

Flash Point Temperature The temperature above which a liquid is capable of producing enough vapor to form a flammable mixture—a flash will occur.

Fundamental Burning Velocity The burning velocity of a laminar flame under stated conditions of composition, temperature and pressure in the unburned gas.

Grounding A procedure of connecting process equipment and materials to an electrical ground in order to bleed off any accumulated electrical charge.

Hazard An inherent physical or chemical characteristic of a material, system, process or plant that has the potential for causing harm.

Hazard Evaluation The analysis of the significance of hazardous situations associated with a process or activity.

Hazard Identification A method used to identify hazards.

Hybrid Mixture A mixture of flammable gas with either a combustible dust or a combustible mist.

Impulse The change in momentum due to a passing blast wave.

Incident The loss of containment of material or energy.

Incident Outcome The physical manifestation of an incident.

Induction Static electricity formation due to a charged object being brought close to an uncharged object.

Induction Time A time delay after ignition of a gaseous mixture before normal flame propagation is observed.

Inherently Safer A chemical process is considered inherently safer if it reduces or eliminates the hazards associated with materials and operations used in the process, and this reduction or elimination is permanent and inseparable.

Inerting An operation involving an inert gas used to achieve a desired fuel or oxygen concentration in order to prevent flammable gas mixtures.

Jet Fire A fire resulting from the combustion of material as it is being released from a pressurized process unit.

Laminar Burning Velocity The speed at which a laminar (planar) combustion wave propagates relative to the unburned gas mixture ahead of it.

LFL *See* Lower Flammability Limit.

Limiting Oxygen Concentration The oxygen concentration below which a fire or explosion is not possible for any mixtures.

Local Ventilation A ventilation method that uses hoods, elephant trunks, or canopies to capture any emitted vapors from a source.

Lower Explosion Limit (LEL) Same as Lower Flammability Limit.

Lower Flammability Limit (LFL) A fuel concentration below which combustion is not possible—the fuel concentration is too lean.

Maximum Experimental Safe Gap (MESG) The maximum clearance between two parallel metal surfaces that has been found, under specified test conditions, to prevent an explosion in a test chamber from being propagated to a secondary chamber containing the same gas or vapor at the same concentration.

Minimum Igniting Current (MIC) Ratio The ratio of the minimum current required from an inductive spark discharge to ignite the most easily ignitable mixture of a gas or vapor, divided by the minimum current required from an inductive spark discharge to ignite methane under the same test conditions.

Minimum Ignition Energy (MIE) The minimum amount of thermal energy released at a point in a combustible mixture that will cause indefinite flame propagation away from that point, under specified test conditions.

Mist Suspended liquid droplets produced by condensation of vapor into liquid or by the breaking up of a liquid into a dispersed state by splashing, spraying or atomizing.

Overpressure The pressure over ambient that results from an explosion.

Padding An inerting method used to provide a continuously maintained atmosphere that is either inert or fuel rich in the vapor space of a container or vessel.

Peak Dynamic Pressure The maximum value of the dynamic pressure.

Peak Overpressure The maximum pressure in a blast or shock wave.

Physical Explosion An explosion due to the sudden release of mechanical energy that does not involve a chemical reaction.

Pool Fire A fire due to surface burning of a flammable or combustible liquid.

Propagating Brush Discharge A static electricity discharge occurring between a grounded conductor and a charged insulator which is backed by a conductor.

Pressure Piling An increase in pressure within a process due to a deflagration. The pressure wave moves ahead of the reaction front, compressing the unreacted gas and increasing the reaction rate of the following reaction front.

Pressure Purging An inerting method that uses a pressurized source of inert gas to reduce the vapor concentration in a process to one that is not combustible.

Propagating Reaction A reaction which propagates spatially through the reaction mass, such as the combustion of a flammable vapor in a pipeline.

Purging An operation involving an inert gas used to achieve a desired fuel or oxygen concentration in order to prevent flammable gas mixtures.

Rapid Phase Transition Explosion An explosion that occurs when a material is exposed to a heat source, causing a rapid phase change and resulting change in material volume.

Reflected Overpressure The pressure measured facing toward the oncoming shock or blast wave.

Runaway Reaction A reaction that occurs when the heat released by the reaction exceeds the heat removal, resulting in a temperature and pressure increase.

Scenario A description of the events that result in an accident or incident.

Shock Wave A pressure front with a very abrupt pressure change.

Side-on Overpressure The pressure due to a shock or blast wave measured at right angles to the shock or blast wave.

Siphon Purging An inerting method that uses a liquid to fill a vessel or process to reduce the vapor concentration to one that is not combustible.

Spark A static electricity discharge that occurs between two conductors.

Sprinkler System A network of piping an discharge nozzles throughout a structure or process through which water is discharged during a fire.

Static Blanketing An inerting method that uses a continuously supplied pressurized source of inert gas to reduce the vapor concentration in a process to one that is not combustible.

Static Electricity Electricity generated by a number of charge mechanisms, including separation, transport, etc.

Streaming Current Static electricity charge generation due to flowing liquids.

Sweep Purging An inerting method that uses a flow of inert gas to reduce the vapor concentration in a process to one that is not combustible.

Thermal Runaway *See* Runaway Reaction.

Transport A static electricity formation mechanism due to the deposition of charged droplets on an object.

Turbulent Burning Velocity A burning velocity that exceeds the burning velocity measured under laminar conditions to a degree depending on the scale and intensity of turbulence in the unburned gas.

Uniform Reaction A reaction that occurs uniformly through space in a reaction mass, such as a reaction that occurs in a CSTR.

UFL *See* Upper Flammability Limit.

Upper Explosion Limit (UEL) Same as Upper Flammability Limit.

Upper Flammability Limit (UFL) A fuel concentration above which combustion is not possible—the mixture is too rich in fuel.

Vacuum Purging An inerting method that uses a vacuum and a source of inert gas to reduce the oxygen concentration in a process such that the vapors are not combustible.

Vapor Cloud Explosion (VCE) An explosion which occurs when a large quantity of flammable vapor or gas is released, mixes with air, and is ignited.

VCE *See* Vapor Cloud Explosion.

Vessel Rupture Explosion An explosion that occurs when a process vessel containing a pressurized material fails suddenly.

INDEX

Publications Available from the
CENTER FOR CHEMICAL PROCESS SAFETY
of the
AMERICAN INSTITUTE OF CHEMICAL ENGINEERS
3 Park Avenue, New York, NY 10016-5991

CCPS Guidelines Series

Guidelines for Process Safety in Outsourced Manufacturing Operations
Guidelines for Process Safety in Batch Reaction Systems
Guidelines for Chemical Process Quantitative Risk Analysis, 2nd Edition
Guidelines for Consequence Analysis of Chemical Releases
Guidelines for Pressure Relief and Effluent Handling Systems
Guidelines for Design Solutions for Process Equipment Failures
Guidelines for Safe Warehousing of Chemicals
Guidelines for Postrelease Mitigation in the Chemical Process Industry
Guidelines for Integrating Process Safety Management, Environment, Safety, Health, and Quality
Guidelines for Use of Vapor Cloud Dispersion Models, Second Edition
Guidelines for Evaluating Process Plant Buildings for External Explosions and Fires
Guidelines for Writing Effective Operations and Maintenance Procedures
Guidelines for Chemical Transportation Risk Analysis
Guidelines for Safe Storage and Handling of Reactive Materials
Guidelines for Technical Planning for On-Site Emergencies
Guidelines for Process Safety Documentation
Guidelines for Safe Process Operations and Maintenance
Guidelines for Process Safety Fundamentals in General Plant Operations
Guidelines for Chemical Reactivity Evaluation and Application to Process Design
Tools for Making Acute Risk Decisions with Chemical Process Safety Applications
Guidelines for Preventing Human Error in Process Safety
Guidelines for Evaluating the Characteristics of Vapor Cloud Explosions, Flash Fires, and BLEVEs
Guidelines for Implementing Process Safety Management Systems
Guidelines for Safe Automation of Chemical Processes
Guidelines for Engineering Design for Process Safety
Guidelines for Auditing Process Safety Management Systems
Guidelines for Investigating Chemical Process Incidents
Guidelines for Hazard Evaluation Procedures, Second Edition with Worked Examples
Plant Guidelines for Technical Management of Chemical Process Safety, Revised Edition
Guidelines for Technical Management of Chemical Process Safety
Guidelines for Process Equipment Reliability Data with Data Tables
Guidelines for Safe Storage and Handling of High Toxic Hazard Materials
Guidelines for Vapor Release Mitigation

CCPS Concept Series

Understanding Explosions
Essential Practices for Managing Chemical Reactivity Hazards
Deflagration and Detonation Flame Arresters
Making EHS an Integral Part of Process Design
Revalidating Process Hazard Analyses
Electrostatic Ignitions of Fires and Explosions
Evaluating Process Safety in the Chemical Industry
Avoiding Static Ignition Hazards in Chemical Operations
Estimating the Flammable Mass of a Vapor Cloud
RELEASE: A Model with Data to Predict Aerosol Rainout in Accidental Releases
Practical Compliance with the EPA Risk Management Program
Local Emergency Planning Committee Guidebook: Understanding
    the EPA Risk Management Program Rule
Inherently Safer Chemical Processes: A Life-Cycle Approach
Contractor and Client Relations to Assure Process Safety
Understanding Atmospheric Dispersion of Accidental Releases
Expert Systems in Process Safety
Concentration Fluctuations and Averaging Time in Vapor Clouds

Proceedings and Other Publications

Center for Chemical Process Safety International Conference and Workshop: Making
    Process Safety Pay—The Business Case
Center for Chemical Process Safety International Conference and Workshop: Process
    Industry Incidents—Investigation Protocols, Case Histories, Lessons Learned,
    2000
Proceedings of the International Conference and Workshop on Modeling the
    Consequences of Accidental Releases of Hazardous Materials, 1999
Proceedings of the International Conference and Workshop on Reliability and Risk
    Management, 1998
Proceedings of the International Conference and Workshop on Risk Analysis in
    Process Safety, 1997
Proceedings of the International Conference and Workshop on Process Safety
    Management and Inherently Safer Processes, 1996
Proceedings of the International Conference and Workshop on Modeling and
    Mitigating the Consequences of Accidental Releases of Hazardous Materials,
    1995
Proceedings of the International Symposium and Workshop on Safe Chemical
    Process Automation, 1994
Proceedings of the International Process Safety Management Conference and
    Workshop, 1993
Proceedings of the International Conference on Hazard Identification and Risk
    Analysis, Human Factors, and Human Reliability in Process Safety, 1992
Proceedings of the International Conference and Workshop on Modeling and
    Mitigating the Consequences of Accidental Releases of Hazardous Materials,
    1991
Safety, Health and Loss Prevention in Chemical Processes: Problems for
    Undergraduate Engineering Curricula